Lecture Notes in Mechanical Engineering

For further volumes:
http://www.springer.com/series/11236

About this Series

Lecture Notes in Mechanical Engineering (LNME) publishes the latest developments in Mechanical Engineering—quickly, informally and with high quality. Original research reported in proceedings and post-proceedings represents the core of LNME. Also considered for publication are monographs, contributed volumes and lecture notes of exceptionally high quality and interest. Volumes published in LNME embrace all aspects, subfields and new challenges of mechanical engineering. Topics in the series include:

- Engineering Design
- Machinery and Machine Elements
- Mechanical Structures and Stress Analysis
- Automotive Engineering
- Engine Technology
- Aerospace Technology and Astronautics
- Nanotechnology and Microengineering
- Control, Robotics, Mechatronics
- MEMS
- Theoretical and Applied Mechanics
- Dynamical Systems, Control
- Fluid Mechanics
- Engineering Thermodynamics, Heat and Mass Transfer
- Manufacturing
- Precision Engineering, Instrumentation, Measurement
- Materials Engineering
- Tribology and Surface Technology

Mohamed Slim Abbes · Jean-Yves Choley
Fakher Chaari · Abdessalem Jarraya
Mohamed Haddar
Editors

Mechatronic Systems: Theory and Applications

Proceedings of the Second Workshop on Mechatronic Systems JSM'2014

Springer

Editors
Mohamed Slim Abbes
Fakher Chaari
Abdessalem Jarraya
Mohamed Haddar
National School of Engineers of Sfax
Sfax
Tunisia

Jean-Yves Choley
SUPMECA
Institut Supérieur de Mécanique
Saint-Ouen
France

ISSN 2195-4356 ISSN 2195-4364 (electronic)
ISBN 978-3-319-07169-5 ISBN 978-3-319-07170-1 (eBook)
DOI 10.1007/978-3-319-07170-1
Springer Cham Heidelberg New York Dordrecht London

Library of Congress Control Number: 2014941629

Printed on acid-free paper

Springer is part of Springer Science+Business Media (www.springer.com)

Contents

Introduction

Mechatronics combines all fields of mechanical and electrical engineering. Mechatronic systems, especially those with a high level of functional integration between mechanical systems, electronic and computer control, have become more and more important for industrial applications. They are employed in various fields, including power systems, transportation, optical telecommunications, and biomedical engineering.

This book includes 15 selected papers presented during the second edition of the Workshop on Mechatronic Systems. This event, held from March 17 to 19, 2014, in Mahdia, Tunisia, was organized within the framework of cooperation between the LAboratory of Mechanics, Modeling and Production (LA2MP) of the National School of Engineers Sfax, Tunisia, and the Laboratory Engineering of Mechanical Systems and Materials (LISMMA) of SUPMECA, in Paris, France.

The workshop provided an excellent forum where researchers from the two laboratories discussed and exchanged the latest advances in mechatronic systems, and presented their recent work and findings on the topic. The first two papers in the book deal with theoretical aspects of mechatronics, including some optimization issues and a topological approach for modeling complex mechatronic systems. The other papers present several applications of mechatronic systems, including considerations on the structural and dynamic behavior of machines. Among those applications, the use of mechatronic systems on gearboxes and the use of milling machines as rotating machines or as part of mechatronic systems highlight the importance of taking into account machine dynamic behavior in the process of designing mechatronic systems or implementing them. Other issues for mechatronic systems are dedicated to production. Collaborative networks partners and their influence on costs reduction are discussed. Products costs and quality level are investigated as objective functions to be optimized in multi-site and multi-plants manufacturing network. In this context, implementation of dedicated software is shown through practical case studies.

Multidisciplinary Optimization of Mechatronic Systems: Application to an Electric Vehicle

Amir Guizani, Moncef Hammadi, Jean-Yves Choley, Thierry Soriano, Mohamed Slim Abbes and Mohamed Haddar

Abstract Preliminary design of mechatronic systems is an extremely important step in the development process of multi-disciplinary products. The great challenge in mechatronic design lies in the multidisciplinary optimization of a complete system with various physical phenomena related to interacting heterogeneous subsystems. In this chapter we combine model-based technique using modeling language Modelica with multidisciplinary optimization approach using Model-Center framework for integrated modeling, simulation and optimization of mechatronic systems. This approach has been applied to the preliminary design of an electric vehicle. Modeling language Modelica has been used to model and simulate the electric vehicle and ModelCenter has been used for the multidisciplinary optimization of the electric motor and the transmission gear ratio. The presented integrated approach allows designers to integrate EV performance analysis with multidisciplinary optimization for efficient design verification and validation.

Keywords Integrated design · Multidisciplinary optimization · Mechatronic systems · Electric vehicle · Modelica · ModelCenter

A. Guizani (✉) · M. Hammadi · J.-Y. Choley · T. Soriano
LISMMA, SUPMECA-PARIS, 3 Rue Fernand Hainaut, Saint-Ouen 93400, France
e-mail: amir.guizani@live.fr

M. Hammadi
e-mail: moncef.hammadi@supmeca.fr

J.-Y. Choley
e-mail: jean-yves.choley@supmeca.fr

T. Soriano
e-mail: thierry.soriano@supmeca.fr

A. Guizani · M. S. Abbes · M. Haddar
LA2MP, ENIS-SFAX, Route de Soukra km 4, 3038 Sfax, Tunisia
e-mail: slim.abbes@enis.rnu.tn

M. Haddar
e-mail: mohamed.haddar@enis.rnu.tn

M. S. Abbes et al. (eds.), *Mechatronic Systems: Theory and Applications*,
Lecture Notes in Mechanical Engineering, DOI: 10.1007/978-3-319-07170-1_1,

1 Introduction

Mechatronic systems (MS) are considered as complex systems composed of many different components. Modeling, simulation and multidisciplinary optimization (MDO) of MS are extremely important steps in the design process (Craig 2009). Each component needs to be properly modeled in order to prevent wrong results and usage. The design or rating of each component is a difficult task as the parameters of one component can affect the power level of another one. There is therefore a risk that one component is inappropriately rated which might make the system unnecessary expensive or inefficient.

Several multi-domain modeling tools such as Bond-Graphs, VHDL-AMS, Matlab/Simulink and Modelica are currently used for preliminary design of MS. For instance, Modelica (Elmqvist et al. 1998) combines object-oriented concepts with multi-port methods for modeling and simulation of physical systems. It includes a declarative mathematical description of models and provides a graphical modeling approach. Multi-domain model library of lumped parameter elements can be created and added to the default Modelica library for future use. The end results of Modelica modeling approach are a system of differential-algebraic equations (DAE) that represents the complete mechatronic system (Hammadi et al. 2012a). So that, Modelica is considered as an ideal tool for preliminary design of MS. However Modelica language has limitations to integrate MDO of MS. One idea is to integrate Modelica language with MDO techniques.

ModelCenter (Long et al. 2008), developed by Phoenix Integration, is a software package that aids in the design and optimization of systems. It enables users to conduct trade studies, as well as optimize designs. It interfaces with other popular modeling tools, including Satellite Tool Kit, Matlab, Nastran and Microsoft Excel. ModelCenter has also tools to enable collaboration among design team members.

Combining ModelCenter with Modelica language allows mechatronic engineers to respond to an important need of integrated design of MS. For this reason, we propose to combine Modelica with ModelCenter to model and optimize an electric vehicle (EV). Optimization is carried out using algorithms available in ModelCenter libraries, especially the Nondominated Sorting Genetic Algorithm (NSGA II) (Deb et al. 2000).

This chapter is organized as follows: After the introduction, we present an analysis of existing solutions. Next, we describe the modeling of the EV. Then, we provide the simulation results of the EV. After that, the optimization of the EV is considered in order to demonstrate the efficiency of the developed models and the approach employed. Finally, we retrieve a conclusion.

2 Analysis of Existing Methods

Several approaches have been developed to solve the multidisciplinary design optimization problem such as concurrent subspace optimization (Braun and Kroo 1995), collaborative optimization CO (Sobieszczanski-Sobieski et al. 1998), multidisciplinary feasible design (Cramer et al. 1994), bi-level integrated system synthesis (Michelena et al. 1999), analytical target cascading ATC (Tappeta and Renaud 1997) and multi objective collaborative optimization MOCO (Choudhary 2004).

These approaches must be chosen and applied by the engineer, according to his knowledge of the problem and his skills. The functioning of these methods can vary greatly. For example Multi-Disciplinary Feasible Design (MDFD), considered to be one of the simplest methods, consists only in a central optimizer taking charge of all the variables and constraints sequentially, but gives poor results when the complexity of the problem increases (Yi et al. 2008). Other approaches, such as Collaborative Optimization or Bi-Level Integrated System Synthesis, are said bi-level. They introduce different levels of optimization (Alexandrov and Lewis 2002), usually a local level where each component is optimized separately and a global level where the optimizer tries to reduce discrepancies among the disciplines. However, these methods can be difficult to apply since they often require to heavily reformulating the problem (Perez et al. 2004) and can have large computation time. To reduce the computing time for the optimization of EVs (Hammadi et al. 2012b) proposed to combine surrogate modeling technique with Modelica language.

3 Electric Vehicle Modeling

3.1 Force Model

The forces which the electric machine of the vehicle must overcome are the forces due to gravity, rolling resistance, aerodynamic drag and inertial effect. The forces acting on the vehicle are shown in Fig. 1.

The total effort of resistance to progress that must defeat the propulsion system to accelerate the vehicle is given by (Janiaud 2011):

$$\begin{aligned} F_t &= F_a + F_r + F_g + F_I \\ &= \frac{1}{2}\rho_{air} \cdot C_{drag} \cdot A_{front} \cdot V_{car} + C_{rr} \cdot M_{car} \cdot g + M_{car} \cdot g \cdot \sin_\alpha + M_{car} \cdot \dot{V}_{car} \end{aligned} \quad (1)$$

where

$F_t[N]$	Traction force of the vehicle
$F_I[N]$	Inertial force of the vehicle
$F_r[N]$	Rolling resistance force

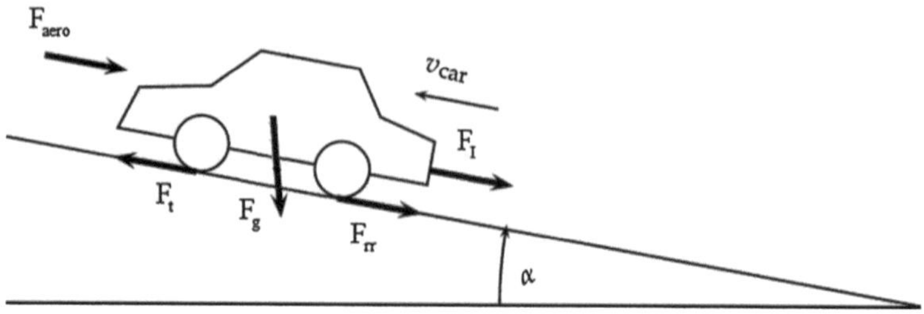

Fig. 1 Free body diagram of the forces (*thick arrows*) acting on the car

$F_g[N]$	Gravitational force of the vehicle
$F_a[N]$	Aerodynamic force
$\dot{V}_{car}\left[m/s^2\right]$	Acceleration of the vehicle
$V_{car}[m/s]$	Velocity of the vehicle
$\alpha[rad]$	Angle of the driving surface
$M_{car}[kg]$	Mass of the vehicle
$g\left[m/s^2\right]$	Free fall acceleration
$\rho_{air}\left[kg/m^3\right]$	Dry air density at 20 °C
$A_{front}[m^2]$	Frontal area
$C_{drag}[-]$	Aerodynamic drag coefficient
$C_{rr}[-]$	Tire rolling resistance coefficient

3.2 Transmission

The torque, angular velocity, and power of the transmission system are given by the following equations (Janiaud 2011):

$$\begin{aligned} \tau_t &= F_t . r_w \\ w_w &= \frac{V_{car}}{r_w} \\ P_t &= F_t . V_{car} \end{aligned} \qquad (2)$$

where

τ_t [Nm]	Traction torque
$w_w[rad/s]$	Angular velocity of the wheels
$r_w[m]$	Wheel radius
$P_t[W]$	Traction power

The traction chain studied here is composed of a mechanical gearbox which connects the motor to the wheels' drive shaft. The gear is characterized by a reduction ratio Gr.

$$\tau_m = \frac{\tau_t}{Gr}$$
$$w_m = Gr.w_w \tag{3}$$

where

τ_m[Nm]	Engine Torque
w_m[rad/s]	Rotation speed of the motor
Gr[-]	Gear ratio of differential

3.3 Electrical Machine: DC Motor

The functional diagram of the electric machine is represented in Fig. 2.

The electric machine is divided into an electric part and mechanic part (Krishnan 2001).

The electric part of the DC-motor is modeled by

$$\mathrm{U(t) = e(t) + R \cdot i(t) + L \cdot \frac{di(t)}{dt}} \tag{4}$$

The mechanical part of the DC-motor can be modeled as follows:

$$\mathrm{J \cdot \frac{dw_m}{dt} = \tau_m(t) - \tau_f(t) - \tau_r(t)} \tag{5}$$

The coupling between the electric and mechanic part is given by

$$\begin{aligned} \mathrm{e(t)} &= \mathrm{K_e \cdot w_m(t)} \\ \mathrm{\tau_m(t)} &= \mathrm{K_c \cdot i(t)} \\ \mathrm{\tau_f(t)} &= \mathrm{a \cdot w_m(t)} \end{aligned} \tag{6}$$

where

U[V]	Armature voltage
i[A]	Current in the armature
R[Ω]	Resistance of the armature
L[H]	Inductance of the armature
e[V]	Electromotive force
J[Kg.m^2]	Moment of inertia
τ_f[Nm]	Friction torque
τ_r[Nm]	Resisting torque

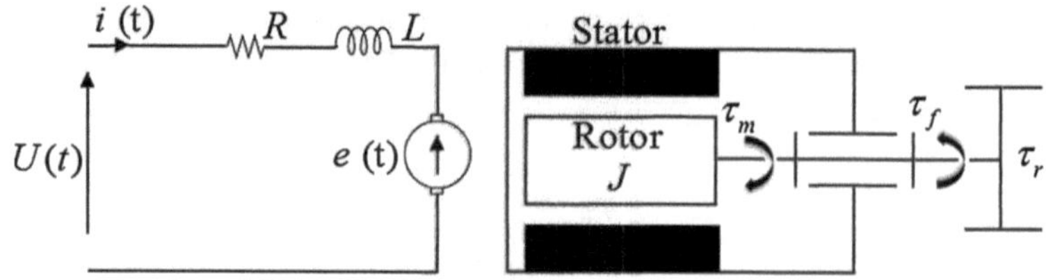

Fig. 2 The functional diagram of DC Motor

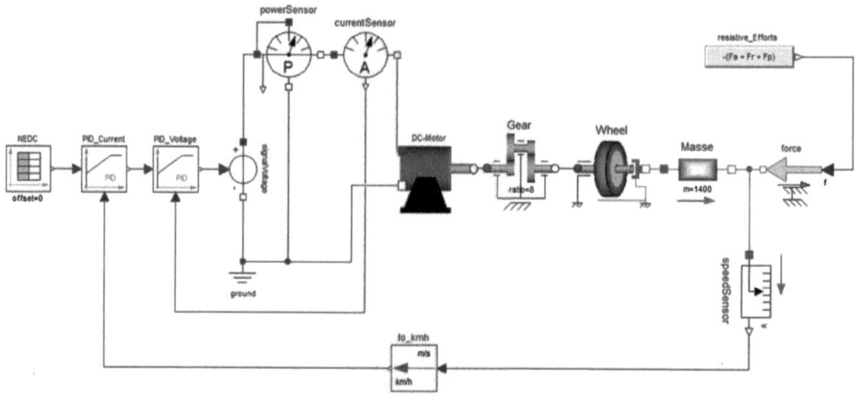

Fig. 3 Electric vehicle model performed by Modelica

K_e[V/rad/s]	Constant emf
K_c[Nm/A]	Torque constant
a[Nm/rad/s]	Coefficient of viscous friction

3.4 Modelica Modeling of the Electric Vehicle

Figure 3 shows the EV model performed using Modelica language. The EV model is composed of the following components: a step input model for velocity demand, a controller, a power sensor, a current sensor, an electric motor and a transmission.

The controller, shown in Fig. 4, is made of the association of two PID controllers (PID_current and PID_voltage), to control current and voltage, respectively. PID_current and PID_voltage have output limitations to define a maximal current Imax and a maximal voltage Umax. The controller input is connected to the velocity demand block.

A simple model of the electric motor is presented by Fig. 5. This model is used to evaluate the electrical power required by the EV. The motor model is composed of a resistance, an inductance and an electromotive component, to convert

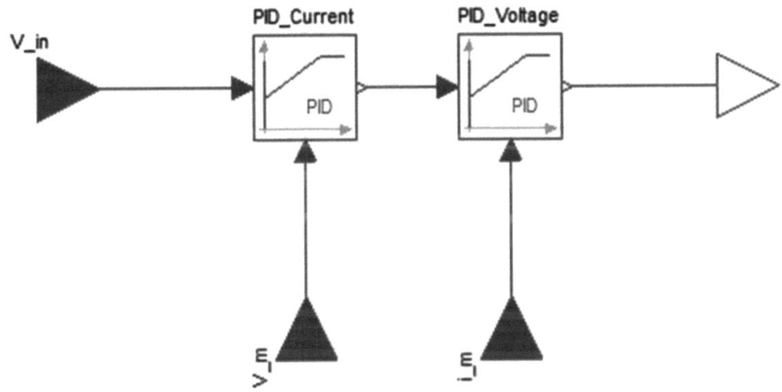

Fig. 4 Controller

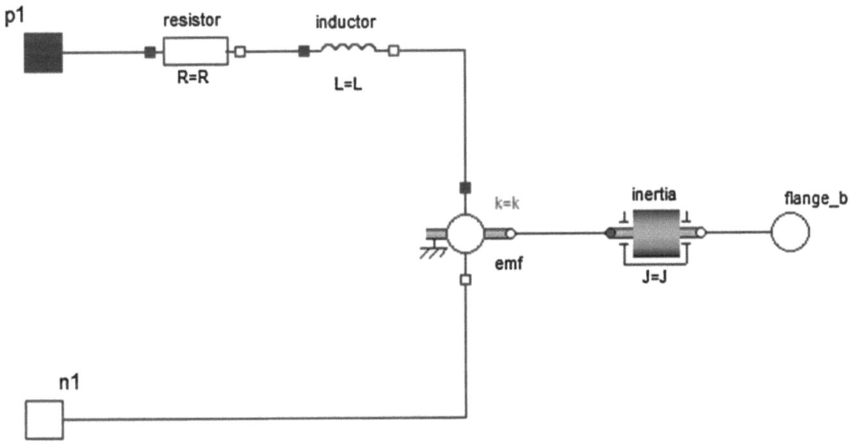

Fig. 5 Electric motor model

electrical power to mechanical power. The inertia component is added to model the motor shaft.

The transmission model is shown by Fig. 6. It is composed of an ideal gear block and a component to model the wheels supporting a tractive force F_t. The vehicle velocity V (km/h) corresponds to the transmission model output.

4 Simulation Results

In the framework of this study, a driving cycle expresses the speed evolution of the vehicle according to the time. It allows evaluating the parameter variations of the vehicle (current of the voltage source, power of the voltage source, tractive force, torque and rotational speed of the drive shaft, etc.).

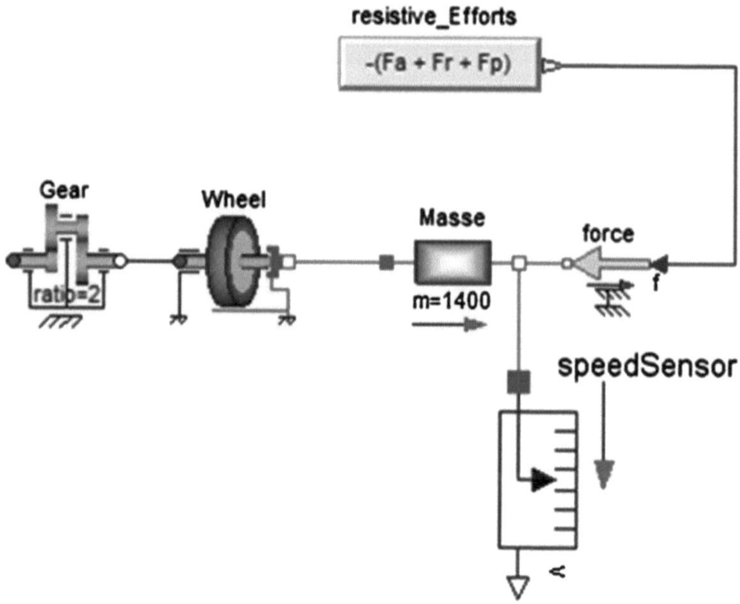

Fig. 6 Transmission model

For the sake of simulation and reproducing a road path with different driving conditions, we will be using the New European Driving Cycle NEDC (Eddine 2010).

In this study, the electromotive force EMF and the transmission gear ratio Gr are supposed known. The objective is to simulate the proposed model. Parameters used in the EV model are given in Table 1.

The simulation parameters of the vehicle according to the NEDC cycle are given by the Figs. 7, 8 and 9.

Figure 7 shows the input signal NEDC and the vehicle speed measured at the output of the component in km/h. This figure confirms that the measured velocity (red curve) follows the profile of the road (blue curve) with an error <2 %.

Figures 8 and 9 show the variation of the electric current and power during the performance test of NEDC. The electric current reaches a maximum of 280 A and the electric power reaches a maximum of 28 kW. The value of the maximum current helps in the choice of the battery.

5 Electric Vehicle Optimization

In this study, we have combined Modelica modeling language with ModelCenter for optimization of an EV. Optimization is carried out using algorithms available in ModelCenter libraries, especially a NSGA II.

Table 1 Electric vehicle parameters

Parameters	Value
Transmission parameters	
Mass of the vehicle M_{car} [Kg]	1400
Free fall acceleration g [m/s^2]	9.81
Dry air density at 20 °C ρ_{air} [Kg/m^3]	1.204
Frontal area A_{front} [m^2]	1.8
Aerodynamic drag coefficient C_{drag}	0.013
Tire rolling resistance coefficient C_{rr}	0.2
Angle of the driving surface α [rad]	0
Radius of the tire r_w [m]	0.3
Gear ratio Gr	8
Electric motor parameters	
Internal resistance R [Ω]	0.02
Inductance L [H]	0.01
Inertia of the motor I [kg/m^2]	0.03
Motor EMF [Nm/A]	0.3
Parameters to simulate	
Speed vehicle V [km/h]	variable
Electric current I [A]	variable
Electric power P [W]	variable

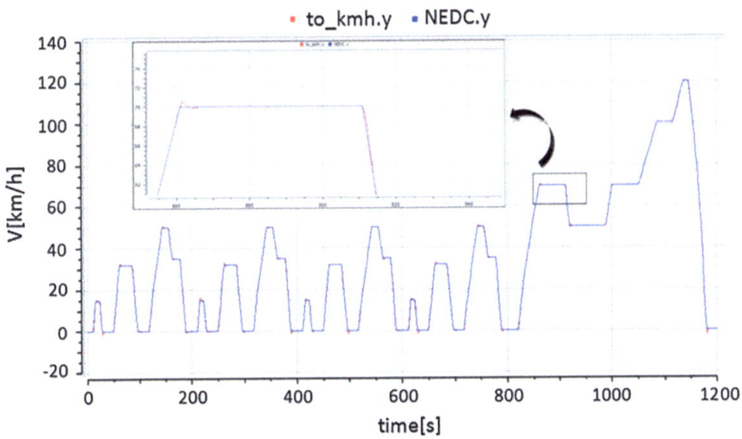

Fig. 7 Speed vehicle for NEDC input

NSGA II algorithm is adapted for multi-objective non-linear optimizing problems. Instead of finding the best design, NSGA tries to find a set of best designs (e.g., Pareto set). A design is said to be dominated if there is another design that is superior to the design in all objectives. The Pareto set consists of non-dominated designs, which are all best in a sense. The Pareto set shows trade-off among competing objectives.

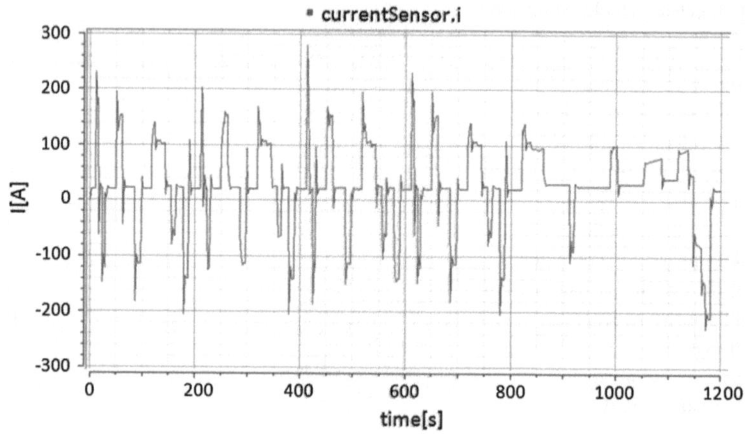

Fig. 8 Electric Current for NEDC

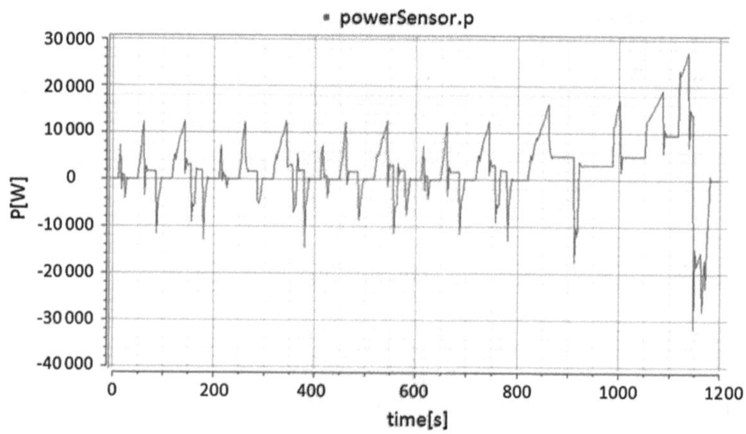

Fig. 9 Electric Power for NEDC

In this study, we are interested in optimizing the road EV in the following cases:

- horizontal road (grade = 0 %).
- 30 % grade road.

The performance requirements needed for the design of a road EV are defined in Table 2.

Based on the precedent mathematical formulation, a Modelica model of the EV has been elaborated to be simulated on an interval time of 100 s. Before starting the optimization, the NEDC cycle is replaced by a constant input V_{max} = 120 km/h (maximum speed of the vehicle).

Table 2 Performance requirements of a road electric vehicle

	V_{10} (km/h): desired velocity at time t = 10 s	V_{max}(km/h): maximal velocity	Power consumption
Optimization 1 (horizontal road)	$V_{10} = 110$	$V_{20} = 120$	Should be minimized
Optimization 2 (30 % grade road)	$V_{10} = 40$	$V_{40} = 80$	Should be minimized

Table 3 Definition of optimization problems

	Design variables	Subject to constraints	Objectives functions
Horizontal road	$2 \leq Gr \leq 15$	$98 \leq V_{10} \leq 102$	min I_{max}
	$0.1 \leq EMF \leq 1$	$117 \leq V_{max} \leq 122$	min U_{max}
30 % grade road	$2 \leq Gr \leq 15$	$38 \leq V_{10} \leq 42$	min I_{max}
	$0.1 \leq EMF \leq 1$	$80 \leq V_{max} \leq 85$	min U_{max}

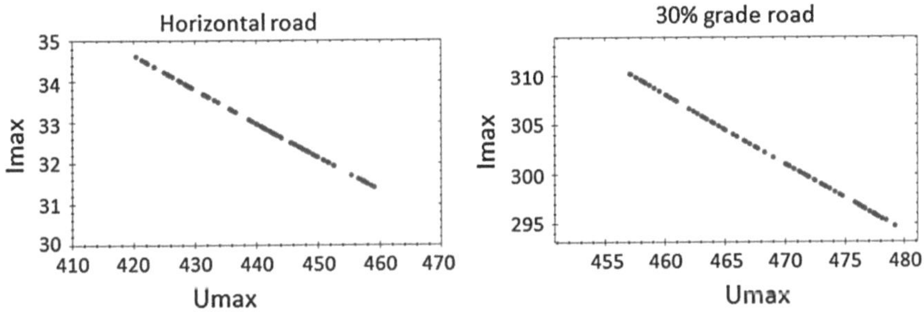

Fig. 10 Pareto front (total solution) for two optimization problems

The input design vector X for ModelCenter is defined as:

$$X = [Gr, EMF]$$

The output variables are defined with V_{10}, V_{max}, I_{max} and U_{max} which are determined by the Modelica simulations at every input design point.

$$Y = [V_{10,}\ V_{max,}\ I_{max,}\ U_{max}]$$

Optimization problems are formulated in Table 3.

Since the optimization is multi-objective, these problems have not a unique solution, but an ensemble of solutions (Pareto Front) given by the Fig. 10.

Each point of the Pareto front is characterized by an input vector X (design variables to optimize) and an output vector (objectives functions and constraints to be respected).

Table 4 Optimizing results using Modelica and ModelCenter

Intervals of the optimal solutions		
Parameters	Horizontal road	Thirty percentage grade road
U_{max}	419.87–459.94	456.89–479.46
I_{max}	31.29–34.65	295.22–310.32
V_{10}	109.12–109.97	38.91–41.50
V_{max}	119.69–120.06	82.43–84.77
EMF	0.25–0.57	0.31–0.67
Gr	6.52–12.49	8.60–14.26

Table 5 Components of X_i

	Gr	EMF
X_1	8.71	0.44
X_2	10.22	0.38
X_3	11.5	0.34

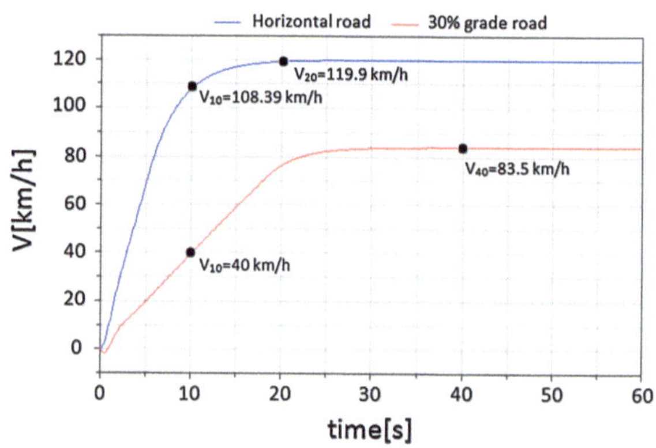

Fig. 11 Speed variation of VE for Gr = 11.5 and EMF = 0.34 V

Table 4 shows the intervals of the optimal solutions for these optimization problems.

The objective of this section is to determine the vector X that satisfies both requirements Y1 and Y2 of the two optimization problems.

The results in Table 4 restrict the search space of design variables as follows: Gr = [8.6, 12.49] and EMF = [0.31, 057].

Based on these two areas of research, the comparison of the Pareto fronts determines the set of X_i vectors which satisfies all the requirements requested.

X_i components are shown in Table 5.

Figure 11 shows the vehicle speed measured at the output of the component to_km/h for X_3 = [11.5, 0.34].

This figure confirms the constraints to be respected in the case of an horizontal slope (optimization 1), the EV reaches a speed of 109.39 km/h in 10 s and a maximum speed V_{max} = 120 km/h in 20 s.

With the same values of EMF and Gr, both constraints in the case of a slope of 30 % (optimization 2) are respected as V_{10} = 40 km/h and $V_{max} = V_{40}$ = 84.3 km/h.

The braking system is not considered in our modeling that's why in the case of a slope of 30 % the speed starts with negative values during a period t = 1 s (this is the response time between the start and the accelerator pedal pressing).

6 Conclusion

In this chapter, an approach combining Modelica and ModelCenter is applied to EV modeling and optimization. Modelica has been used to model the EV and ModelCenter has been used to optimize design variables. Results show that the proposed approach allows designers to integrate easily complex models with performance analysis and multidisciplinary optimization.

Thus the proposed approach helps designers of MS in verification and validation of the mechatronic design and therefore in making decisions efficiently.

Acknowledgements The authors would like to thank the researchers of the laboratory LISMMA in SUPMECA-Paris, who have contributed in this work with their helpful comments and suggestions.

References

Alexandrov N, Lewis R (2002) Analytical and computational aspects of collaborative optimization for multidisciplinary design. AIAA J 40(2):301–309

Braun RD, Kroo IM (1995) Development and application of the collaborative optimization architecture in multidisciplinary design environment, multidisciplinary design optimization: state-of-the-art. Proceedings of Applied Mathematics, SIAM, Philadelphia

Choudhary R (2004) A hierarchical optimization framework for simulation-based architectural design, PhD Thesis, University of Michigan

Craig K (2009) Mechatronic system design. In: Proceedings of the Motor, Drive and Automation Systems Conference

Cramer E, Dennis J, Frank P, Lewis R, Shubin G (1994) Problem formulation for multidisciplinary optimization. SIAM J Optim 4(4):754–776

Deb K, Agrawal S, Pratap A, Meyarivan T (2000) A fast elitist non-dominated sorting genetic algorithm for multi-objective optimization: NSGA-II .Lecture notes in computer science, vol 1917. Springer, Heidelberg, pp. 849–858

Eddine GS (2010) Modélisation, commande et gestion de l'énergie d'un véhicule électrique

Elmqvist H, Mattsson S, Otter M (1998) Modelica: The new object-oriented modeling language. 12th European Simulation Multiconference, Manchester, UK

Janiaud N (2011) Modélisation du système de puissance du véhicule électrique en régime transitoire en vue de l'optimisation de l'autonomie, des performances et des coûts associés. Supélec

Hammadi M, Choley JY, Penas O, Rivière A (2012a) Mechatronic system optimization based on surrogate models-application to an electric vehicle. In: 3rd International conference on simulation and modeling methodologies, technologies and applications (SIMULTECH), pp. 11–16

Hammadi M, Choley JY, Penas O, Rivière A (2012b) Multidisciplinary approach for modelling and optimization of Road Electric Vehicles in conceptual design level. In: IEEE electrical systems for aircraft, railway and ship propulsion (ESARS), 2012, pp 1–6

Krishnan, R (2001) Electric motor drives: modeling, analysis, and control. Prentice Hall, Englewood Cliffs, p 626

Long T, Liu L, Wang J, Zhou S, Meng L (2008) Multi-objective multidisciplinary optimization of long-endurance UAV wing using surrogate models in Model Center. In: Proceedings of the 12nd AIAA/ISSMO multidisciplinary analysis and optimization conference, pp 1–8

Michelena N, Papalambros P, Park H, Kulkarni D (1999) Hierarchical overlapping coordination for large-scale optimization by decomposition. AIAA J 37:890–896

Perez R, Liu H, Behdinan K (2004) Evaluation of multidisciplinary optimization approaches for aircraft conceptual design. In: AIAA/ISSMO multidisciplinary analysis and optimization conference, Albany, NY

Sobieszczanski-Sobieski J, Agte JS, Sandusky R (1998) Bi-level integrated system synthesis (BLISS), Symposium on multidisciplinary analysis and optimization, 7th, Street, Louis, MO, Collection of technical papers. Pt. 3, USA, pp 1543–1557

Tappeta R, Renaud J (1997) Multi-objective collaborative optimization. ASME J Mech Des 119:403–411

Yi S, Shin J, Park G (2008) Comparison of mdo methods with mathematical examples. Struct Multi Optim 35(5):391–402

Topological Approach for the Modeling of Complex and Mechatronic Systems

Mariem Miladi Chaabane, Régis Plateaux, Jean-Yves Choley, Chafik Karra, Alain Riviere and Mohamed Haddar

Abstract The main objective of this chapter is to show the advantage of the application of a topological approach for the modeling of complex and mechatronic systems. This approach is based on the notions of topological collections and transformations and it is applied by using the MGS language (Modeling of General Systems). The various applications studied by applying this approach such as bars and beams structures and elementary mechatronic components (piezoelectric structures and motor reducer) are presented.

Keywords Mechatronic systems · Topological approach · Topological collections · Transformations · KBR graph · MGS language

1 Introduction

Thanks to technological development and because of consumers requirements, new systems called mechatronic systems, appeared. Among these systems we can quote as examples the SEGWAY which is a single-seat electric vehicle with two wheels, the Electronic Wedge Brake (EWB) which replaces a whole hydraulic brake system and the Nao robot which is an intelligent humanoïde robot. These systems have

M. M. Chaabane (✉) · C. Karra · M. Haddar
Mechanical, Modeling and Manufacturing Unit, National Engineering School of Sfax (ENIS), BP 1173, 3038 Sfax, Tunisia
e-mail: mariam.mi@hotmail.fr

M. M. Chaabane · R. Plateaux · J.-Y. Choley · A. Riviere
Laboratory of Engineering of the Mechanical Structures and Materials, High Institute of Mechanic of Paris (SUPMECA), 3, rue Fernand Hainaut, 93407 Saint-Ouen Cedex, France

M. M. Chaabane
Higher Institute of Industrial Systems of Gabes (ISSIG), 6011 Gabes, Tunisia

M. S. Abbes et al. (eds.), *Mechatronic Systems: Theory and Applications*, Lecture Notes in Mechanical Engineering, DOI: 10.1007/978-3-319-07170-1_2,

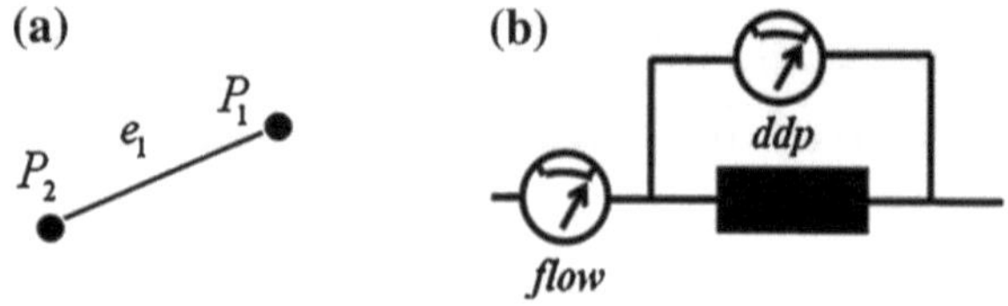

Fig. 1 **a** Topological structure. **b** Physical component

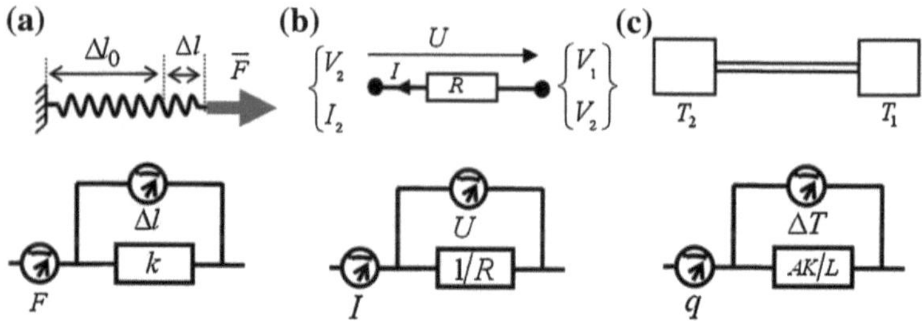

Fig. 2 **a** Spring. **b** Electric resistance. **c** Metal bar

a high level of integration of electronics, mechanics, automatic and data processing and then present complex behaviors also these mechatronic systems present new constraints of precision and size as well as new environmental constraints. From where the need for a new modeling approach allowing to guarantee the continuity and the coherence of the modeling of mechatronic systems (Plateaux 2011; Plateaux et al. 2007). Therefore, in this work we are interested to adopt a topological approach for the modeling of mechatronic systems. In fact, each field of mechatronic can be characterized by its topological structure and its behavior law.

2 Contribution of Topology for the Modeling

Considering the topological structure presented by the Fig. 1a which is a linear graph composed of an arc noted e_1 and two nodes noted P_1 and P_2. By associating variables of type flow and potential difference to this topological structure, it can present various physical components (Fig. 1b). For example, basing on the inverse analogy which retains the topology (Firestone 1933), this topological structure can present a spring in traction compression in the mechanical field (Fig. 2a), an electric resistance in the electrical field (Fig. 2b) or a metal bar in the thermal field (Fig. 2c).

In order to apply this approach, we are based on the topological graph named KBR, in the honor of their creators Kron, Branin and Roth (Fig. 3) (Kron 1963; Branin 1966; Roth 1955). The KBR graph enables the distinguishing between the topological structure of the system and the associated physics and then can be used as a unification basis for the modeling of complex or mechatronic systems.

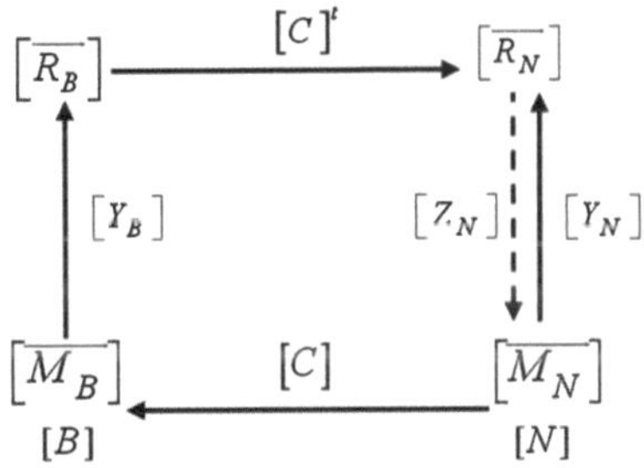

Fig. 3 The KBR topological graph

As a language for the application of the KBR graph, we initially applied MODELICA language. Indeed, MODELICA is an object-oriented language that uses the concept of flow and potential thus respecting the Kirchhoff laws. Also, it offers a great number of free libraries and it has the advantage of being interfaced with other software such as CATIA V6 and MATLAB/Simulink…. On the other hand its topological nature is limited to 0 and 1complexes and the access to higher dimensions can be done only via transformations. For example a variable magnetic field adding a stray current in a control circuit could be taken into account in MODELICA only if the model describes with its connectors (0-simplexes) is created. Also, it associates the topology and the behavior in the same model that limits the generalisation of the studied model. Therefore, the limitations of MODELICA language impose another language for the KBR graph (Plateaux 2011).

In our study, we resorted to use the MGS language which is an abreviation of General Modeling System. This language is a research project in the IBISC (Laboratory for Computer Science, Integrative Biology and Complex Systems) of the University of Evry (France) (Spicher 2006; Cohen 2004). This project study the contribution of topological concepts in the programming languages and apply these concepts to the design of new data structures and control for the modeling and simulation of dynamic systems with dynamic structures. In addition to its basic elements, MGS integrates a new kind of values called topological collections which consist of a set of cells organized with a neighbourhood relationship and decorated by values. To manipulate its data structure, MGS uses the transformations which are defined by a set of rewriting rules of the following form $m \Rightarrow e$. The left-hand part of the rule is called pattern and the right-hand part is the expression that replaces the instances of m. We mainly distinguish functions defined by case, paths transformations which allow the update of values associated with cells and patches which are intended to modify the structure of the cells. Then, we are interested to use the topological collections to present the topology of the studied system which means the interconnection laws between its components and the transformations to specify the local behavior law of these components.

The general modeling approach using topological collections and transformations consists in presenting the studied system by a cellular complex to which we associate the variables of interest. Then, we specify the local behavior law and

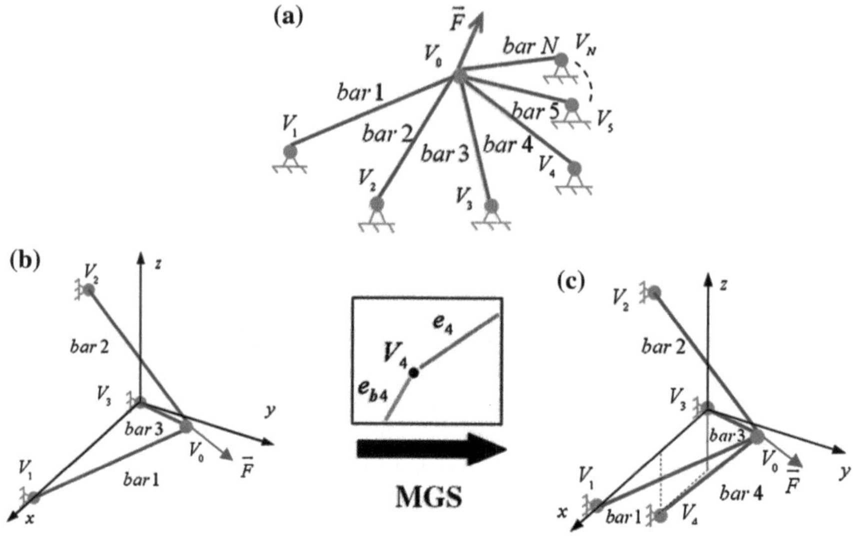

Fig. 4 **a** N bar truss structure with pyramidal structure **b** three bar truss structure **c** four bar truss structure

equilibrium equations of the different components of the studied systems. Finally, the generation of the system of equations is done by sweeping all the cells representing the system. This system is written in MODELICA format and we use DYMOLA as a solver.

3 Applications

In order to validate the topological approach presented in this chapter, we applied it in a first part to the modeling of mechanical structures of bars and beams. Then in a second part, we applied it to the modeling of mechatronic components (Miladi Chaabane 2014).

First of all, we applied this topological approach in the particular case of an N bar truss structure with pyramidal structure (Fig. 4a). Two particular cases are studied which are an isostatic case for a three bar truss structure (Fig. 4b) and a hyperstatic case for a four bar truss structure (Fig. 4c) (Miladi Chaabane et al. 2013a). Using topological collections, the passage from a three bar truss structure to a four bar truss structure is done only by adding the cells related to the addition of the fourth bar and the parameters which are associated to these cells (Fig. 4).

Then, we apply this topological approach to the modeling of plane and space bar structures by studying the particular case of a two-bar plane truss structure (Fig. 5) (Miladi Chaabane et al. 2012) and finally, we generalized this approach to the modeling of plane and space beam structures by studying the case of a plane

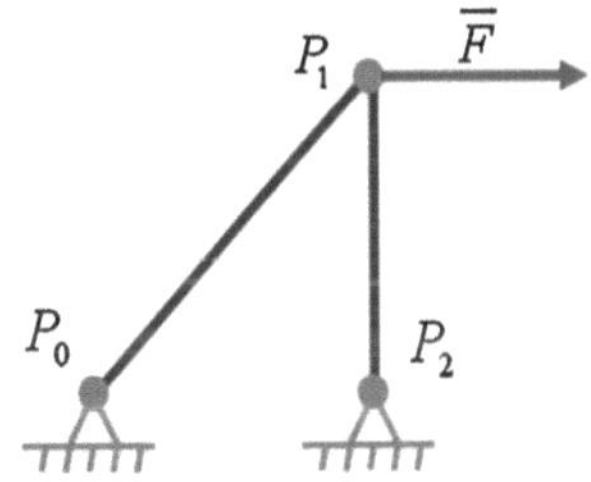

Fig. 5 Application for bar structures: two-bar plane truss structure

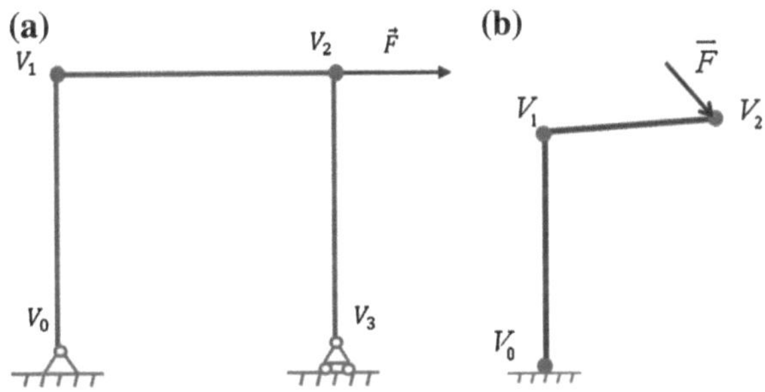

Fig. 6 Application for beam structures **a** plane portal and **b** spatial structure made up of two beams and non-coplanar force acting on it

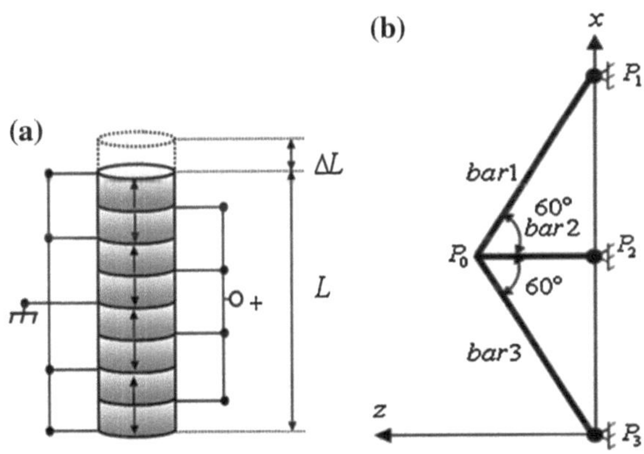

Fig. 7 Application for piezoelectric structures **a** multi layer piezoelectric stack and **b** piezoelectric truss structure

portal (Fig. 6a) and the case of a spatial structure made up of two beams and non-coplanar force acting on it (Fig. 6b).

The advantage of the application of the topological collections and their transformations for the modeling of bar and beam structures compared to the other approaches is that we declare the local behavior law of the bars or the beams independently of their numbers and the way in which they are connected i.e. of their topology. Indeed, we consider a bar or a beam as a local element.

In the second part, the application of the topological collections and their transformations is extended to the modeling of more complex mechanical systems. First of all we applied this approach to the modeling of piezoelectric structures by

Fig. 8 Motor reducer

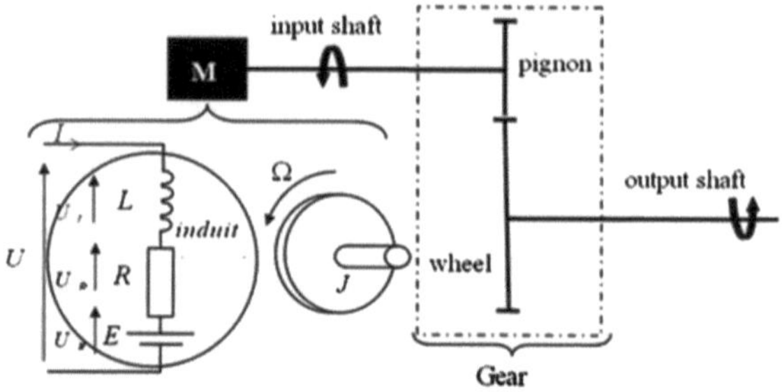

studying the case of a multi layer piezoelectric stack (Fig. 7a) and the case of a piezoelectric truss structure (Fig. 7b) (Miladi Chaabane et al. 2013a, b). Finally we applied the topological collections and their transformations for the modeling of a motor reducer (Fig. 8).

For the various cases presented by Figs. 7 and 8, the studied systems are described by a set of local interactions between elementary entities. For example, using the topological collections, the motor reducer is modeled by taking into account all its elements (resistor, inductor, EMF, inertia, gear (pinion/wheel), input and output shafts) and then the motor reducer is considered as a set of local elements linked by neighborhood relationships.

Indeed in this example, the topological collections are used to present the topological structure of the motor reducer i.e. the interconnection law between its elements and the transformations are used to specify the local behavior law as well as the equilibrium equations of each component of the motor reducer. The generation of the system equation is done by sweeping all the cells representing the motor reducer.

4 Conclusion

In this chapter, we presented a brief overview of the application of a topological approach for the modeling of complex and mechatronic systems on the basis of topological collections and transformations. This topological approach allows the consideration of topological relations. Indeed the topological structure of a system is independent of its behavior. Also, this topological approach allows the simplification of the modeling of complex systems and then a complex system is described by a set of local interactions between elementary entities. Finally, this approach allows taking into account of the multi-scale aspect and then the model associated with the component can be more or less important from a functional or physical point of view.

On the other hand, in the different studied examples, we are limited to the case of mechanical structures and elementary mechatronic systems. However, it is necessary to extend this topological approach for the modeling of systems which integrate all the fields of mechatronic by integrating data processing and the automatic. Also, it would be interesting to apply the topological transformations of type patchs which allow the automatic refinement. Finally, we can create a MGS library by the determination of the local behaviors laws of the various fields of mechatronic systems.

References

Branin FH Jr (1966) The algebraic-topological basis for network analogies and the vector calculus. In: Fox J (ed) Proceedings of the symposium on generalized networks, Wiley-Interscience, New York

Cohen J (2004) Intégration des collections topologiques et des transformations dans un langage fonctionnel. Ph.D. thesis, University of Evry Val d'Essonne

Firestone F (1933) A new analogy between mechanical and electrical systems. J Acous Soc Am 4:249–267. doi:10.1121/1.1915605

Kron G (1963) Diakoptics: the piecewise solution of large-scale systems. MacDonald, London

Miladi Chaabane M, Plateaux R, Choley J-Y et al (2013a) New approach to solve dynamic truss structure using topological collections and transformations. Int J Mech Sys Eng (IJMSE) 3:162–169

Miladi Chaabane M, Plateaux R, Choley J-Y et al (2103b) New topological approach for the modeling of mecatronic systems: application for piezoelectric structures. Eur J Comput Mech. doi:10.1080/17797179.2013.820896

Miladi Chaabane M (2014) Modélisation géométrique et mécanique pour les systèmes méca(tro)niques. PhD thesis, Institut Supérieur de Mécanique de Paris et Ecole Nationale d'Ingénieurs de Sfax

Miladi Chaabane M, Plateaux R, Choley J-Y, Karra C, Riviere A, Haddar M (2012) Topological approach to solve 2D truss structure using MGS language. IEEE MECATRONIC REM. 437–444

Plateaux R, Penas O, Rivière A, Choley J-Y (2007) A need for the definition of a topological structure for the complex systems modeling. CPI'2007—Rabat- Maroc

Plateaux R (2011) Continuité et cohérence d'une modélisation des systèmes mécatroniques basée(s) sur une structure topologique. PhD thesis, Institut Supérieur de Mécanique de Paris

Roth JP (1955) An application of algebraic topology to numerical analysis: on the existence of a solution to the network problem. Proc Natl Acad Sci 41:518–521

Spicher A (2006) Transformation de collections topologiques de dimension arbitraire. Application à la modélisation de systèmes dynamiques. PhD thesis, University of Evry Val d'Essonne

Instrumentation of Back to Back Planetary Gearbox for Dynamic Behavior Investigation

Ahmed Hammami, Alfonso Fernández del Rincón, Fakher Chaari, Fernando Viadero Rueda and Mohamed Haddar

Abstract The objective of this chapter is to investigate experimentally the dynamic behavior of back-to-back two stages planetary gear under stationary and non stationary conditions. Instrumentation layout for measurement is presented. In order to get signal from carriers, a developed instrumentation is showed: connections are achieved between accelerometer and acquisition system through a hollow slip ring. After that, series of measurements are achieved to study the dynamic behavior of carriers under stationary condition and the influence of the run up regime in the dynamic behavior of carriers.

Keywords Planetary gear · Back-to-back · Carriers · Dynamic behavior · Stationary condition · Run up

---

A. Hammami (✉) · F. Chaari · M. Haddar
Laboratory of Mechanics, Modeling and Production, National School of Engineers of Sfax, 1173, 3038 Sfax, Tunisia
e-mail: ahmed_hammami@voila.fr

F. Chaari
e-mail: fakher.chaari@gmail.com

M. Haddar
e-mail: mohamed.haddar@enis.rnu.tn

A. Hammami · A. Fernández del Rincón · F. Viadero Rueda
Faculty of Industrial and Telecommunications Engineering, Department of Structural and Mechanical Engineering, University of Cantabria, Avda de los Castros, s/n, 39005 Santander, Spain
e-mail: alfonso.fernandez@unican.es

F. Viadero Rueda
e-mail: fernando.viadero@unican.es

M. S. Abbes et al. (eds.), *Mechatronic Systems: Theory and Applications*, Lecture Notes in Mechanical Engineering, DOI: 10.1007/978-3-319-07170-1_3,

1 Introduction

Planetary gears are useful in many fields of application for transmitting significant power with large speed reductions or multiplications. The excessive applied torque, the variable speed and manufacturing and/or installation errors cause instability and vibration under stationary or non stationary operating conditions.

In fact, most studies are devoted to stationary condition: (Hidaka and Terauchi 1976). Kahraman (1999) studied the load sharing behaviour of planetary gear set and (Iglesias et al. 2013) studied the effect of planet position error on load sharing and transmission error. Ligata et al. (2008) studied experimentally the influence of the influence of manufacturing errors on the planetary gear stresses and the planet load sharing. Inalpolat and Kahraman (2009) presented a theoretical and experimental investigation on modulation sidebands observed in planetary gear.

The speed's fluctuation will modify the structure of the frequency response: Randall (1982) justify that the vibration amplitude which is induced by meshing process is modulated by load fluctuation. Bartelmus et al. (2010) found that the shape of the transmission error function changes as a result of load variation. Chaari et al. (2011a, b) studied the influence of local damage and time varying load on the vibration response and the influence of variable load which induces the variability of the transmission speed (Chaari et al. 2013). Kim et al. (2012) studied the dynamic behaviour of a planetary gear when component gears have time-varying pressure angles and contact ratios. Hammami et al. (2013) studied the dynamic behaviour of rings of back-to-back planetary gear under stationary and non-stationary conditions.

In order to validate experimentally the numerical results, many researchers developed a different kind of test bench. In order to analyze the vibration of spur gear with variable spacing, Chaari (2005) recorded the signal of the bearing of the motor's shaft with "Kistler" piezoelectric accelerometer and HP 35665A analyzer. Wu et al. (2012) used a simplified test rig with a pair a spur gears, shafts and bearing sets. The accelerometer was PCB Model 355B04, the data acquisition device was NI9234. The encoder measured the rotating speed of the driving shaft is synchronously measured through the encoder. The command of the electrical motor was through the motor control unit and NI9401 DIO card. Schon (2005) used a conventional helical gearbox test bench. Measurement hardware was composed of four ICP accelerometer and "Siglab" data acquisition. Using the same measurement hardware, (Schon 2005) used an epicyclic gearbox test bench driven by an electrical motor.

Boguski (2010) used a planetary test machine to avoid the dynamic effects under quasi-static planetary gear. The driven motor is connected to a torque meter "Lebow 1288–10 K". A torque sensor is connected to the sun gear. Two optical encoders (Heidenhain IBV600) controlled the speed of the sun and the carrier and are connected to the software application Transmission Error Measurement System (TEMS) by Superior Controls, Inc.

For planetary gear Ligata (2007) and Inalpolat (2009) used the same test bench which is a back-to-back planetary gear set configuration: the electric motor powers the suns of both gear set. Ligata used a 32 channel strain gauge conditioner unit (four NI SCXI-1520 strain bridge modules mounted on NI SCXI-1000 chassis), a PC with suitable data acquisition boards (NI PCI-6052E multifunction I/O board), and a magnetic pick-up for identification of a particular planet as it passes through a given gauge location. Inalpolat used eleven accelerometers (PCB-353B15) mounted radially on the outside surface of the ring and a photoelectric sensor measured the angular speed of the sun gear shaft.

In this chapter, we interested in the characterization of dynamic behaviour of a carrier on back-to-back planetary gear. First, a description of the instrumentation used to achieve this target. Then, experiments are achieved under stationary condition. After that, the test bench run in ramp up regime.

2 Description of the Test Bench

A test bench is developed at the department of structural and mechanical engineering of the University of Cantabria in Spain. It is composed of two identical planetary gear sets with the same gear ratio.

2.1 Architecture of the Test Bench

The mechanical parameters of the test bench are given in Table 1.

For economic and energy efficiency criteria, the configuration of the test bench is compact with mechanical power circulation (Fernández et al. 2013). The test bench is composed of two identical planetary gear sets. The first planetary gear set is a test gear set and the second is a reaction gear set. They are connected back-to-back: the sun gears of both planetary gear sets are connected through a common shaft and the carriers of both planetary gear sets are connected to each other through a rigid hollow shaft (Fig. 1).

2.2 Motor Drive Selection

An electric motor is connected to the shaft of the sun gear to rotate both gear sets.

Speed control of the electric motor “SIEMENS” is made by a frequency inverter “MICROMASTER 440” which can be configured with the help of the software “STARTER”. This software, developed by “SIEMENS”, can be started directly through the frequency converter integrated into PCS 7.

Table 1 Mechanical parameters of the planetary gear

	Carrier	Ring	Sun	Planet
Teeth number	–	65	16	24
Moment of inertia (kgm^2)	0.0021	0.697	0.0003	0.002
Base diameter (mm)	57.55	249.38	61.38	92.08

Fig. 1 Back-to-back layout as assembled in the bench

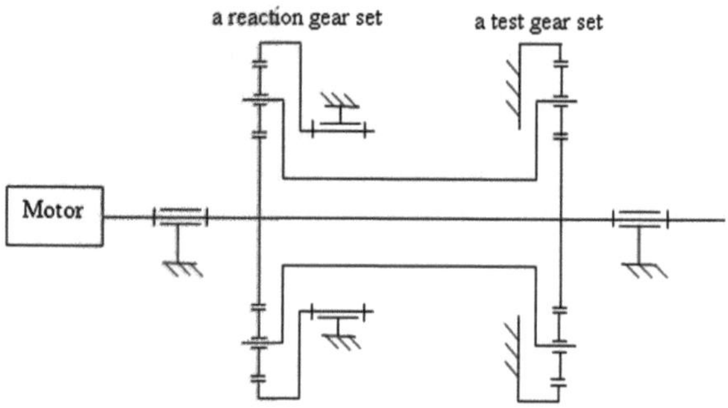

2.3 *Instrumentation and Data Analysis*

2.3.1 Instrumentation of the Test Rig

Four ISOTRON tri axial accelerometers are mounted on the test bench: two on each ring (Figs. 2 and 3).

The Table 2 characterise each ISOTRON accelerometer.

Vibration signals registered by accelerometers will be acquired by a LMS SCADAS 316 system and the data will be processed with the software "LMS Test.Lab" to visualize time history of accelerations, their spectra and time frequency maps. This software allows visualisation of signals under stationary and non-stationary conditions. Fast Fourrier Transform (FFT) is the signal processing method used by this software for analysing stationary signals whereas the Short Time Fourrier Transform (STFT) is the signal processing method used for analysing non-stationary signals and can be showed as waterfall or colormap.

It is possible to select in the "Acquisition setup" of this software the "Span", "the frequency lines" and the "resolution". The span is the frequency range over which the measurement will be taken and which is unaffected by the cutoff filter. It is 80 % of the bandwidth. The frequency lines are the number of lines in the data block. The block size is 2 times the number of lines. The frequency resolution is that of the data block. It is related to the frequency range and the number of lines in the block (Resolution = Bandwidth /Frequency lines).

Additionally, an optic tachometer (Compact VLS7) is combined with pulse tapes along the axe in order to measure its instantaneous angular velocity. It was placed along the hollow shaft.

Fig. 2 Tri axial accelerometers on the reaction ring

Fig. 3 Tri axial accelerometers on the test ring

Table 2 Accelerometer's characteristics

Mounting points	Free ring	Fix ring	Free ring	Fix ring
Serial number	10020	10021	10022	10023
Moment of inertia (kgm^2)	0.0021	0.697	0.0003	0.002
X-Sensitivity (mV/g)	102.6	103.5	100.4	101.6
Y-Sensitivity (mV/g)	101.3	98.68	99.67	103.0
Z-Sensitivity (mV/g)	101.1	104.3	102.6	101.9
Manufacturer	ENDEVCO–ISOTRON			
Model number	65 M–100			

2.3.2 Developed Instrumentation

The acquisition of vibratory signals is strongly affected by the occurrence of modulations at the same spin frequency as the carriers. In order to avoid this phenomenon, a slip ring (HBM SK5/95) with 5 wires is installed with the hollow shaft that connects the carriers and, therefore, it allows the installation of accelerometers on these elements (Fig. 4).

The slip ring body is composed of a hollow cylinder with five hard silver slip rings.

Two brush holders SK5/ZB with five brushes each complete the slip ring body to form the assembly. The five brushes are arranged for signal transmission on the holder. They can move around a common bolt and springs will provide the necessary mechanical pressure. Two holes with M3 thread are performed for the mounting of the brush holder.

The instrumentation layout with the extended slip ring is on Fig. 5.

To connect the accelerometers to the acquisition system "LMS SCADAS 316", we should connect accelerometers to the slip ring body in a first time and then connect the brushes of slip ring to the acquisition system.

The used connections are male and female BNC connector and male and female RS232 ports.

The four first wire of the slip ring are for signal channel whereas the fifth wire which is connected to the hollow shaft is used as common mass.

Figure 6 shows the connection between accelerometers and the slip ring body.

Every brushes of the slip ring have a colour. The white, the black, the blue and the red brushes are for signal channel whereas the yellow brush is used as common mass.

Figure 7 shows the connection between the brushes of the slip ring and the acquisition system "SCADAS LMS 316".

3 Results

Experiments are achieved under stationary condition (a fixed external load and a fixed speed) and under non-stationary condition (a variable speed under run up regime).

First of all, we define in the software of the acquisition system the channels (Fig. 8): the channel group identity, the direction of accelerometers, the input mode, the measured quantity, the measured quantity, the electrical unit, the sensitivity of accelerometers and the sensitivity unit.

After calibration of accelerometer, we have to define in the tracking setup measurement mode if it is stationary or tracking.

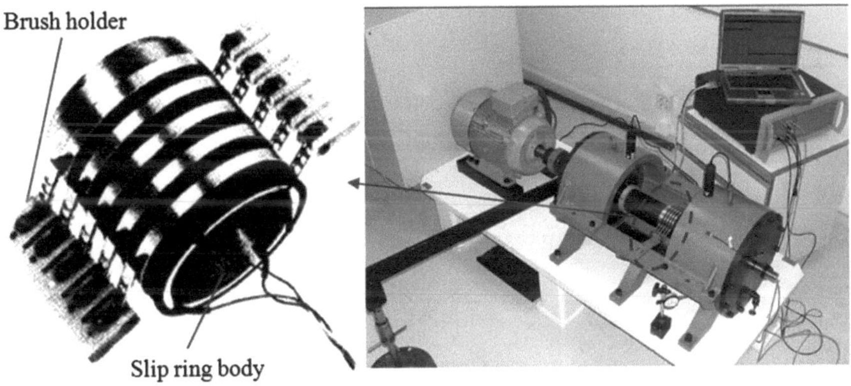

Fig. 4 Slip ring assembly SK5

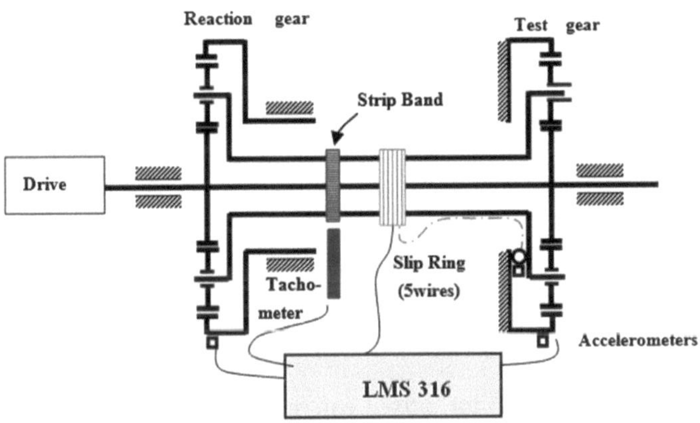

Fig. 5 Schematic representation of the instrumentation layout

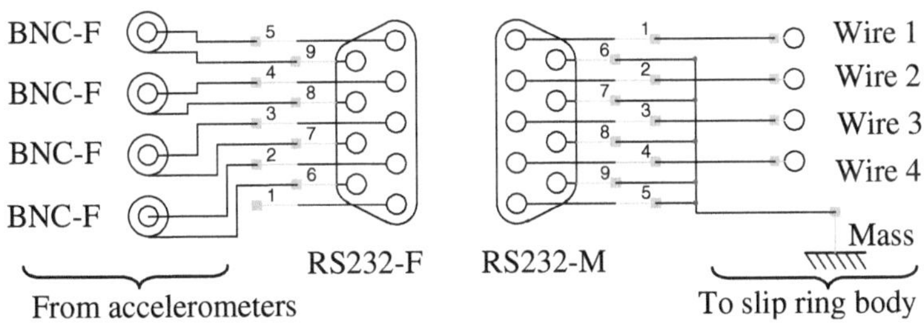

Fig. 6 Connection between accelerometers and the slip ring body

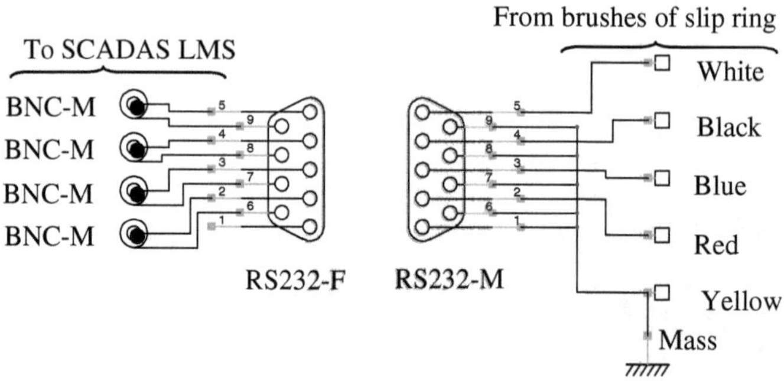

Fig. 7 Connection between brushes of the slip ring and the acquisition system

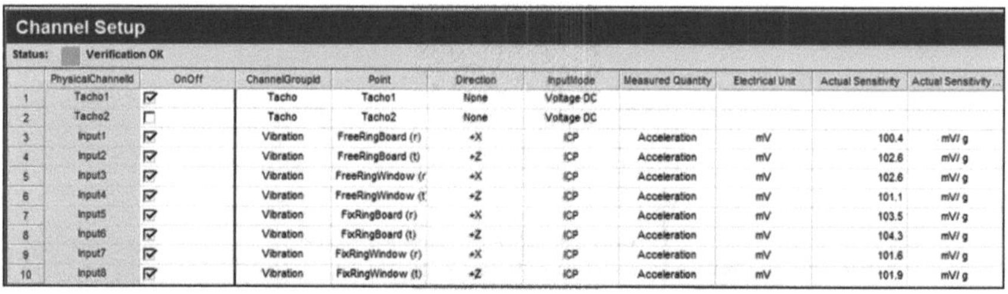

Channel Setup

Status: Verification OK

	PhysicalChannelId	OnOff	ChannelGroupId	Point	Direction	InputMode	Measured Quantity	Electrical Unit	Actual Sensitivity	Actual Sensitivity...
1	Tacho1	☑	Tacho	Tacho1	None	Voltage DC				
2	Tacho2	☐	Tacho	Tacho2	None	Voltage DC				
3	Input1	☑	Vibration	FreeRingBoard (r)	+X	ICP	Acceleration	mV	100.4	mV/ g
4	Input2	☑	Vibration	FreeRingBoard (t)	+Z	ICP	Acceleration	mV	102.6	mV/ g
5	Input3	☑	Vibration	FreeRingWindow (r	+X	ICP	Acceleration	mV	102.6	mV/ g
6	Input4	☑	Vibration	FreeRingWindow (t	+Z	ICP	Acceleration	mV	101.1	mV/ g
7	Input5	☑	Vibration	FixRingBoard (r)	+X	ICP	Acceleration	mV	103.5	mV/ g
8	Input6	☑	Vibration	FixRingBoard (t)	+Z	ICP	Acceleration	mV	104.3	mV/ g
9	Input7	☑	Vibration	FixRingWindow (r)	+X	ICP	Acceleration	mV	101.6	mV/ g
10	Input8	☑	Vibration	FixRingWindow (t)	+Z	ICP	Acceleration	mV	101.9	mV/ g

Fig. 8 Channel setup

3.1 Stationary Conditions

First experiment is achieved for a fixed external load (0 Nm on the free reaction ring) and a fixed speed (1,500 rpm). So, measurement mode is stationary in the "LMS Test Lab" (Fig. 9). We define the duration of measurement, the acquisition rate, number of averages and the averaging type.

In the Acquisition Setup, we define the acquisition parameters: bandwidth, resolution and the frequencies lines which are calculated automatically. We define also the view settings: the window, the spectrum format and the format.

After starting motor, we start ranging accelerometers and finally we measure vibration. Figure 10 shows acceleration in g unit on the test carrier on three period of rotation of carrier ($T_c = 0.2$ s, 1 g $= 9.806$ m/s^2).

In Fig. 10, a clear amplitude modulation is observed with 3 repetitive increasing in amplitude corresponding to the force due to rotation of carrier. In fact, when a planet approaches the accelerometer location, higher vibration levels will be registered.

The spectra of dynamic component of the test carrier (Fig. 11) shows that the back-to-back planetary gears are excited by the rotation of carrier ($f_c = 5$ Hz) meshing frequency ($f_m = 320.7$ Hz) and its harmonics. In addition, sidebands appear on these spectra on the n.f_c (n: integer).

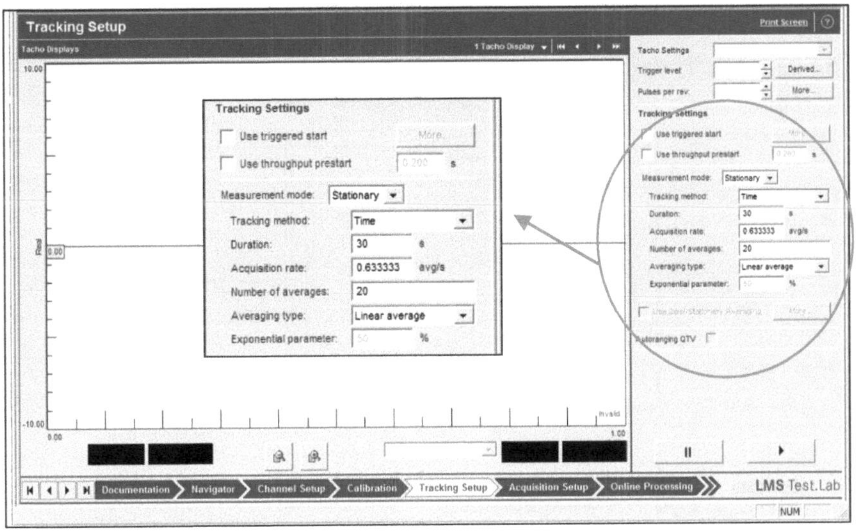

Fig. 9 Tracking setup for stationary condition

Fig. 10 Time response of the test carrier

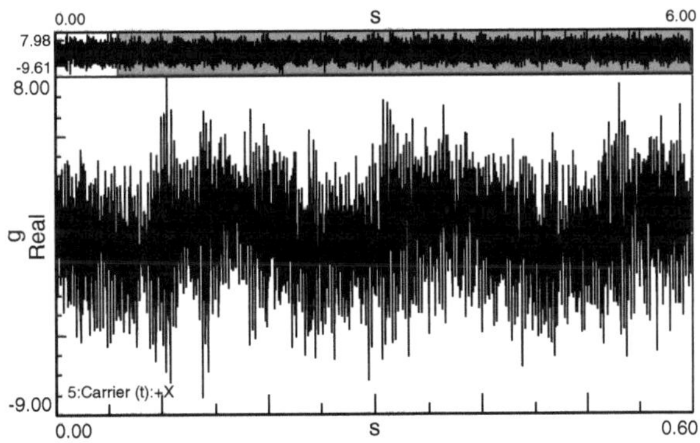

Fig. 11 Frequency response of the test carrier

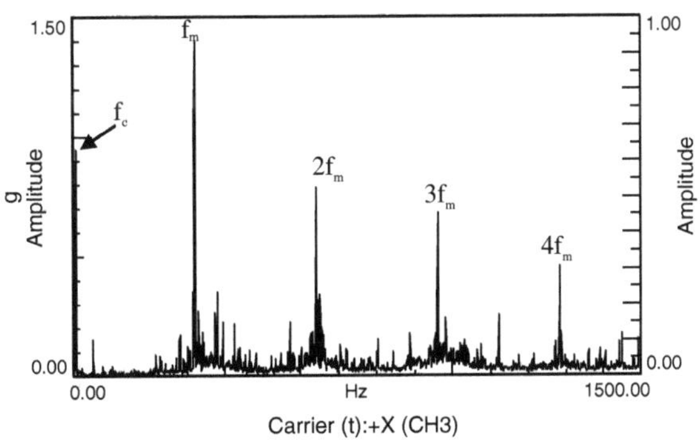

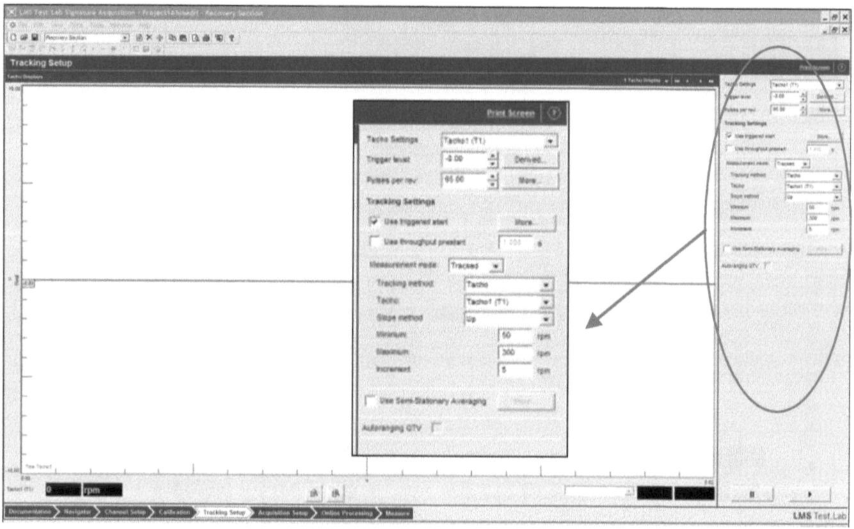

Fig. 12 Tracking setup for run up and run down

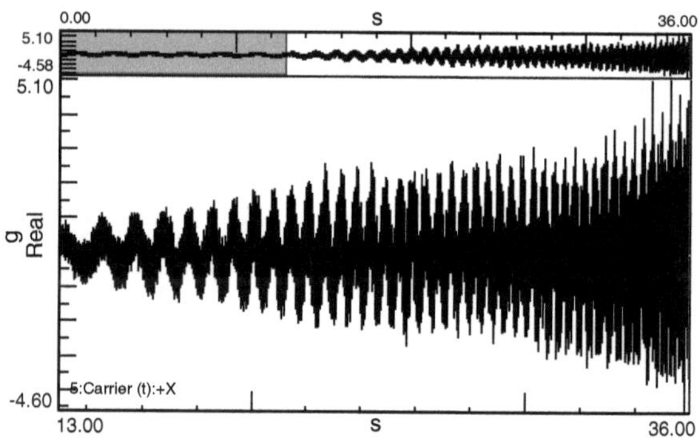

Fig. 13 Time response of the test carrier in the run up

3.2 Run Up

The variation of speed is controlled by the frequency converter Micromaster 440 which commands linearly the variation of the rotational speed of motor in the run up regime. In the "LMS Test Lab" software, we define the tacho settings. The measurement mode will be "Tracked" and the slope method is "up" for run up regime and "down" for run down regime (Fig. 12).

The system is driven from 30 rpm of speed of carrier to 190 rpm during 23 s. The time response of the acceleration on the test carrier is shown in Fig. 13. During run up, the accelerating torque is increasing, giving rise to increased vibration and amplitude of oscillation.

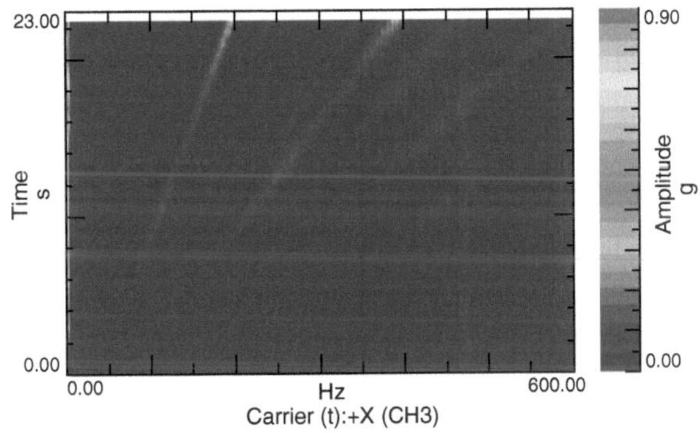

Fig. 14 Color map of the test carrier dynamic response in the run up

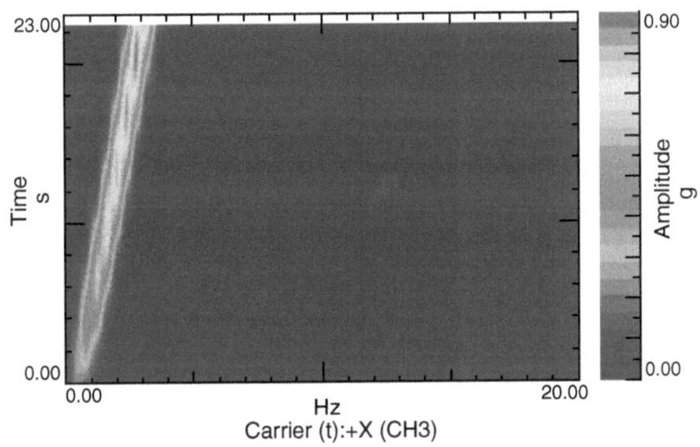

Fig. 15 Color map of the test carrier dynamic response in the run up (zoom)

In order to describe the evolution of the frequency content during the run up, a color map is drawn based on Short Time Fourrier Transform (STFT). Figure 14 shows STFT obtained from experiments for acceleration on the test carrier. The amplitude of acceleration of the test carrier (in the g unit) is presented in the frequency (X-axes) and in the time (Y-axes).

We clearly observe inclined lines showing the increase of the meshing frequency and its harmonics on Fig. 14. In addition, an important acceleration is shown in the low-frequencies. A zoom in 0–20 Hz of Fig. 14 is showed in Fig. 15.

Figure 15 shows an inclined line where the amplitude is more important. This inclined line describes the evolution of the rotation frequency of carriers.

4 Conclusion

A back-to-back planetary gear test bench under fixed speed and variable speed conditions was presented.

A first study considering stationary operation (fixed speed and load) was conducted. An investigation of modulation sidebands of the test carrier were highlighted experimentally.

The second study was with variable speed which is delivered by an asynchronous driving motor and controlled by a frequency converter in the run up regime. The acceleration of the test carrier was amplified by this variation of the speed. Short time frequency analysis was operated in order to characterize frequency content and identify the speed variation. The colormap obtained from experience for acceleration on the test carrier showed inclined lines corresponding to the variation of the meshing frequency and its harmonics and the variation of the rotation of carrier frequency where the amplitude is more important.

Future investigation will be focused mainly on the dynamic behaviour of planets under stationary and non-stationary condition.

Acknowledgements This chapter was financially supported by the Tunisian-Spanish Joint Project N° A1/037038/11.

References

Bàguski BC (2010) An experimental investigation of the system-level behavior of planetary gear sets. Thesis, the Ohio State University

Bartelmus W, Chaari F, Zimroz R, Haddar M (2010) Modelling of gearbox dynamics under time-varying non stationary load for distributed fault detection and diagnosis. Eur J Mech A/Solids 29:637–646

Chaari F (2005) Contribution to the study of the dynamic behaviour of planetary gear set with defects. Dissertation, National Engineering school of Sfax, University of Sfax, Sfax

Chaari F, Zimroz R, Bartelmus W, Fakhfakh T, Haddar M (2011a) Modelling of planetary gearbox for fault detection. Investigation on local damage size and time-varying load conditions influence to vibration response. In: The eighth international conference on condition monitoring and machinery failure prevention technologies

Chaari F, Zimroz R, Bartelmus W, Fakhfakh T, Haddar M (2011b) Modelling of local damages in spur gears and effects on dynamics response in presence of varying load conditions, Surveillance 6 conference (Oct 2011)

Chaari F, Abbes MS, Viadero F, Fernandez A, Haddar M (2013) Analysis of planetary gear transmission in non-stationary operations. Front Mech Eng 8(1):88–94

Fernández A, Cerdá R, Iglesias M, De-Juan A, García P, Viadero F (2013) Test Bench for the analysis of dynamic behavior of planetary gear transmissions. In: The 3rd international conference on condition monitoring of machinery in non-stationary operations (CMMNO 2013)

Hammami A, Fernández A, Chaari F, Viadero F, Haddar M (2013) Dynamic behaviour of two stages planetary gearbox in non-stationary operations. Surveillance 7 conference (Oct 2013)

Hidaka T, Terauchi Y (1976) Dynamic behavior of planetary gear-1st report, Load distribution in planetary gear. Bull Japan Soc Mech Eng 19:690–698

Iglesias M, Fernandez A, De Juan A, Sancibrian R, Garcia P (2013) Planet position errors in planetary transmission: effect on load sharing and transmission. Front Mech Eng 8(1):80–87

Inalpolat M (2009) A theoretical and experimental investigation of modulation sidebands of planetary gear sets. Dissertation, the Ohio State University

Inalpolat M, Kahraman A (2009) A theoretical and experimental investigation of modulation sidebands of planetary gear sets. J Sound Vib 323:677–696

Kahraman A (1999) Static load sharing characteristics of transmission planetary gear sets: model and experiment, transmission and driveline systems symposium SAE paper 1050
Kim W, Lee JY, Chung J (2012) Dynamic analysis for a planetary gear with time-varying pressure angles and contact ratios. J Sound Vib 331:883–901
Ligata H (2007) Impact of system-level factors on planetary gear set behaviour. Dissertation, the Ohio State University
Ligata H, Kahraman A, Singh A (2008) An experimental study of the influence of manufacturing errors on the planetary gear stresses and planet load sharing. J Mech Des 130–041701:1–9
Randall RB (1982) A new method of modelling gear faults. J Mech Des 104(2):259–267
Schon PP (2005) Unconditionally convergent time domain adaptive and time-frequency techniques for epicyclic gearbox vibration. Thesis. The University of Pretoria
Wu TY, Chen JC, Wang CC (2012) Characterization of gear faults in variable rotating speed using Hilbert-Huang transform and instantaneous dimensionless frequency normalization. Mech Syst Signal Process 30:103–122

Experimental Study of Bolted Joint Self-Loosening Under Transverse Load

Olfa Ksentini, Bertrand Combes, Mohamed Slim Abbes, Alain Daidié and Mohamed Haddar

Abstract Self-loosening of bolted joints is a problem regularly encountered in aeronautical structures and much research has been devoted to this phenomenon. Developing detailed analytical equations for the dynamic study of unscrewing is difficult, so it is easier to reveal it experimentally, in the static or the dynamic case. This paper focuses on the experimental study of the dynamic self-loosening of a bolted joint when it is subjected to vibrations, a major cause of the problem. The experiment used a bolted assembly moved by a shaker, which caused the tightened parts to slide and the bolt to loosen. A load washer showed the axial load variation and a high speed camera recorded the movements of the assembled parts. The results show the progress of rotation of the different parts, the unscrewing of the bolt and the loss of tension in the assembly. The method provides a means to explore the loosening process of various types of bolts, under realistic conditions of vibration.

Keywords Self-loosening · Bolt joint · Transverse load · Experimental study

O. Ksentini (✉) · M. S. Abbes · M. Haddar
Mechanic Modeling and Production Laboratory LA2MP, National School of Engineers of Sfax, BP 1173-3038 Sfax, Tunisia
e-mail: Olfa.ing@gmail.com

M. S. Abbes
e-mail: slim.abbes@enis.rnu.tn

M. Haddar
e-mail: mohamed.haddar@enis.rnu.tn

O. Ksentini · B. Combes · A. Daidié
Institute Clement Ader, National Institute of Applied Sciences of Toulouse, University of Toulouse, 31077 Toulouse, France
e-mail: bcombes@insa-toulouse.fr

A. Daidié
e-mail: alain.daidie@insa-toulouse.fr

M. S. Abbes et al. (eds.), *Mechatronic Systems: Theory and Applications*, Lecture Notes in Mechanical Engineering, DOI: 10.1007/978-3-319-07170-1_4,

1 Introduction

Many structures are assembled with bolt joints since they are easily removable. However, problems such as fatigue and self-loosening are encountered when this system is used in mechanical and aeronautical structures.

The connecting bolt is usually made with irreversible threads and the risk of self-loosening can occur only under the influence of an external load (torque or force) that can overcome the friction load, or when there is a significant reduction of the friction due to relative sliding of the contacting surfaces. Nevertheless, this problem is still not well controlled since it has many possible causes, such as sizing errors, vibration, and variations of temperature.

Several techniques can be used to prevent this problem, e.g. deformable washers, but their diversity reflects the complexity of the problem and the difficulty of solving it. Many authors have addressed the issue experimentally, analytically or numerically and have shown that the physical phenomena differ depending on the load direction.

Due to the complexity of dynamic problems, most of this research has been limited to quasi static and static cases.

Several studies have focused on the self-loosening problem. Early research concentrated on bolt joint self-loosening when it was subject to an axial load. Goodier and Sweeney (1945) developed an experiment to highlight the self-loosening phenomenon. It consisted of a traction machine for loading and unloading a screw nut system. The relative rotation of the nut was measured using a microscope. They tested a bolt joint subjected to dynamic load. Despite their inability to achieve complete bolt joint loosening, they offered an explanation of the partial loosening of fasteners.

Hess (1998) and Hess and Sudhirkashyap (1997) conducted experimental tests to observe the influence of several parameters, such as the pretension, the thread pitch and the external load, on self loosening. Their experiment consisted of applying a compressive load to a bolt by means of a cam.

Aziz (2003), among others, studied bolt joint self-loosening under axial load. In his experiment, an inertial mass was excited by a shaker. This led to separation of the nut from its support surface, but without unscrewing. Thus he concluded that a properly sized assembly cannot be loosened when it is subjected to axial load.

So how is it possible for a bolted joint to loosen under a transverse load? Many authors have attempted to answer this question. They have concluded that the most commonly encountered self-loosening is produced when the displacement of the clamped parts is sufficient to cause sliding under the nut or the screw head. In the literature, the most widely used apparatus is the Junker machine (Junker 1969). This machine consists of a motor that drives an eccentric mechanism causing the displacement and sliding of one of assembled part. A load sensor is used to measure the preload and an accelerometer measures the acceleration.

A second approach, used by Dong and Hess (2000) to achieve the unscrewing, is the shock test. In this experiment, tubes are tightened by means of a bolt and move in slots of a part subjected to a reciprocating movement. At each vibration cycle, a shock occurs in the bolt when the tube abuts on the extremities of the hole. The shock causes movement of the washer, which can lead to slipping under the nut or the screw head and thus to loosening of the bolt.

Bhattacharya et al. (2010) is among the authors who have studied the ways of preventing the loosening of a bolt joint, and the energy dissipated by friction between two assembled parts has been studied by Bouchaala et al. (2013).

In addition to experiments, there are many analytical and numerical works on the subject in the literature. Dinger and Friedrich (2011) has proposed a finite element model to simulate bolt self-loosening. He then compares experimental results to the numerical ones, basing his experiment on the Junker machine. Analytical approaches have been developed by Nassar and Yang (2009), who presents mathematical models to study bolted joint self-loosening and compares his results with experimental tests. His work is also based on the Junker machine.

Each author has developed a strategy to reveal the bolt joint loosening experimentally. Except for Aziz, all authors have used the Junker machine, in which rolling elements are added at the interface between assembled parts. But in real applications, there is no rolling element at this interface. In order to study the self loosening problem rigorously, we developed an experimental model with direct friction at the interface between assembled parts.

The objective of this chapter is to develop an experiment to analyse the dynamic behaviour of the bolted joint when it is subjected to an external load acting in the transverse direction. The theory of this experiment is presented and then some results are given.

2 Experiment

To obtain bolt joint loosening, a transverse dynamic load was generated to overcome the frictional load under the nut or under the screw head so that the nut and/or the screw slipped.

The experimental set-up is shown in Fig. 1; it consisted mainly of a mass clamped by a bolt and moved by a shaker.

The different loads are summarized in Fig. 2.

The underlying theory is that an inertial load leads to the sliding of the mass on the bracket. So the screw bends under the mass sliding and generates shear loads that can lead to a transverse sliding of the head or of the nut. Secondly, the bolt preload generates pressure on the thread flank, which generates a torque in the loosening direction under the effect of the helix angle. This torque is insufficient by itself to overcome friction and to cause sliding. But when the screw slips transversally, it is deflected by the thread torque, so it rotates slightly while moving transversally. It can also be the nut which rotates. In both cases, self-loosening occurs.

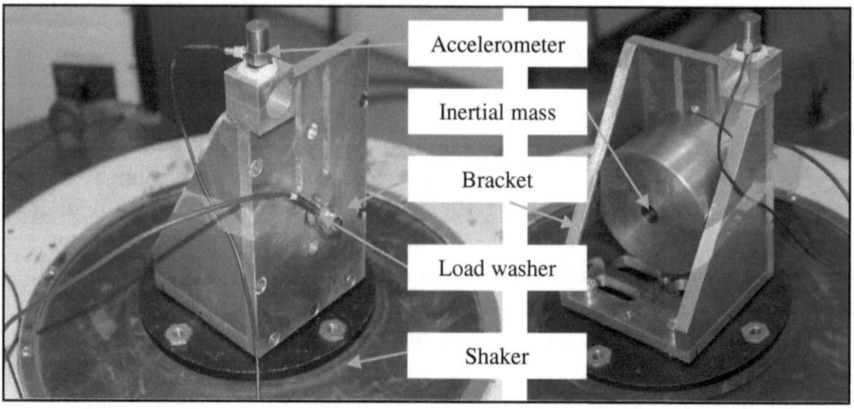

Fig. 1 Experimental set-up

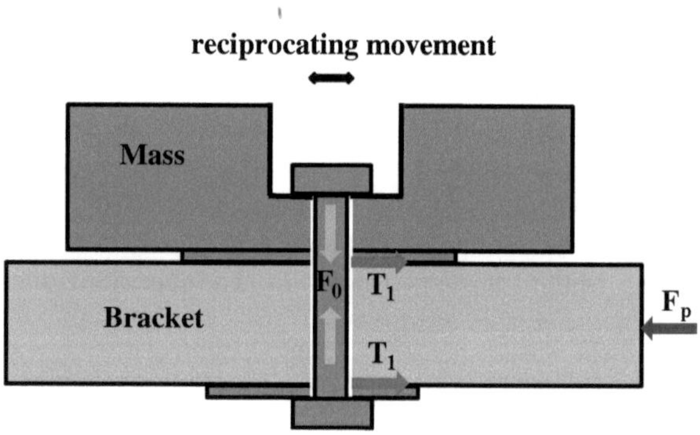

Fig. 2 Loads on the experimental device

The mass and the bracket were not chosen arbitrarily; the steps used are summarized in Fig. 3.

In the first step, in order to choose a preload, a clamping ratio representing a percentage of screw stress with respect to the yield strength of the screw material was fixed. Then, based on Eqs. (1) and (2), all friction loads were calculated. In the fourth step and for the chosen masses, the acceleration necessary to obtain sliding was estimated. Finally the shaker load was calculated based on Eq. (3):

$$\mathrm{T}_1 = f_1 F_0 \quad (1)$$

$$a = \frac{2\mathrm{T}_1}{m_i} \quad (2)$$

$$\mathrm{F_p} = \left(m_p + m_i + m_e\right)a \quad (3)$$

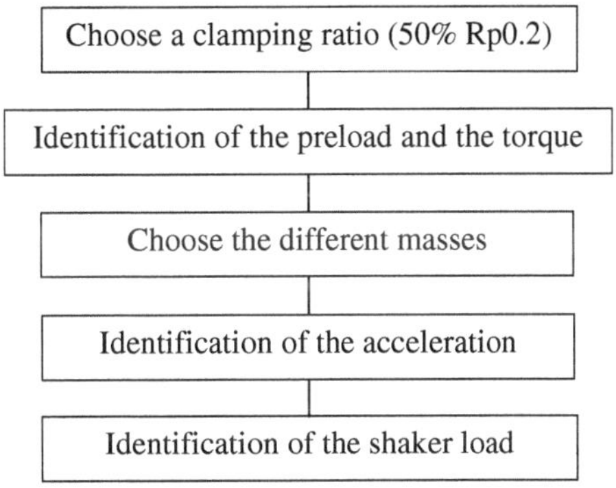

Fig. 3 Flowchart of successive steps

with:

T_1 the friction force,
f_l the friction coefficient,
F_0 the tension in the bolt,
a the overall acceleration coming from the shaker movement,
m_i the inertial mass,
F_p the force required from the shaker,
m_p the shaker mass,
m_e the bracket mass.

Generally, the normal stress on a tightened bolt varies between 50 and 70 % of the yield stress (Rp0.2). But since we were limited by the shaker capacities, the clamping ratio used was only 50 %. So, for a M6 bolt of class 8.8 and a friction coefficient of 0.15, the weight of the inertial mass and the bracket were respectively 3 and 1.2 kg and the shaker load was equal to 5 kN.

Figure 4 shows all the measuring equipment.

The load washer KMR20 was provided by HBM. It consists of strain gauges and is designed for the measurement of static and dynamic compressive strength. It is especially suitable for monitoring efforts in machinery production. This sensor was placed between the nut and the bracket to monitor the screw axial load variation according to time.

It was difficult to detect the rotation of the nut or the screw with the naked eye, so a high speed camera was used to take a large number of pictures per second. Indicator needles were placed on the screw and the nut to facilitate the rotation measurement.

The assembly was caused to vibrate in the transverse direction. All tests were performed at a fixed frequency of 52 Hz with a progressive increase in the vibration amplitude. The shaker amplitude was measured by an accelerometer.

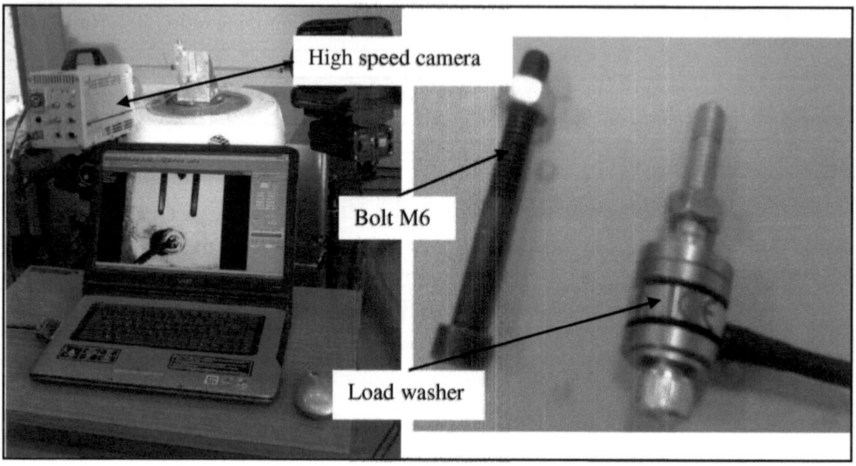

Fig. 4 Measuring equipment

3 Results

First we show the results obtained in the case of a treated screw. Then we discuss the influence of the screw type on the dynamic behaviour of the assembly.

Figure 5 shows the variation of the axial load measured by the load washer for a treated screw. Two steps can be seen. At first, despite the increasing acceleration, there is no real variation of the axial load. The increase of the measure is explained by the tilting induced in the cantilevered mass.

When the inertial load becomes higher than the friction forces, the screw slips and rotates in the loosening direction. The mass can rotate in the screwing or the unscrewing direction but, for the majority of tests, it rotated in the screwing direction, which could lead to the increase in the preload.

The decrease in the preload was evidence of the loosening phenomenon.

The fast camera provided 50 images per second. Needles placed on the screw and the nut helped to show when and where the loosening occurred. Most results showed that the loosening resulted from the rotation of the screw. So, at the beginning of this phenomenon, there was rotation of the screw in the loosening direction without any rotation of the nut.

Figure 6 shows rotation of the screw and the nut. There are two phases of interest in this figure: the first one corresponds to the rotation of the screw in the screwing direction, which leads to the increase in the preload. The second step corresponds to the beginning of the loosening phenomenon, in which the screw and the nut rotate in the unscrewing direction.

We can conclude that self-loosening results mainly from the sliding of the parts. The mass slips under the bracket, which causes the screw to bend. At first there is no sliding under the screw head. But when the mass sliding is sufficient it causes sliding under the head.

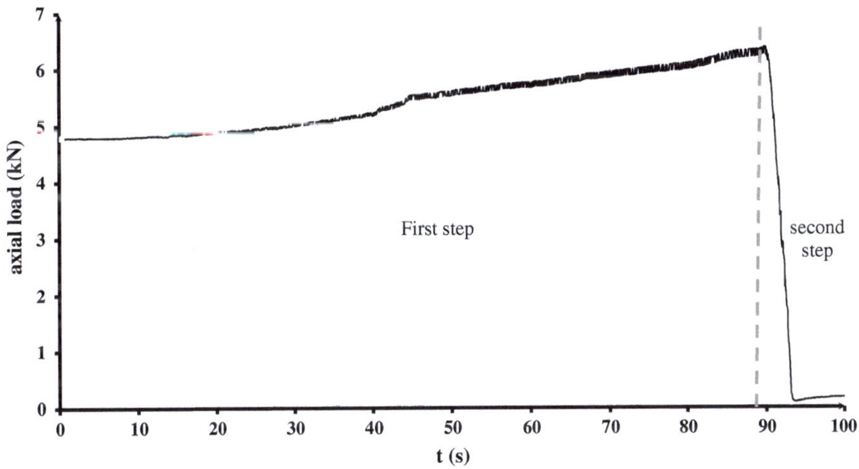

Fig. 5 Axial load variation

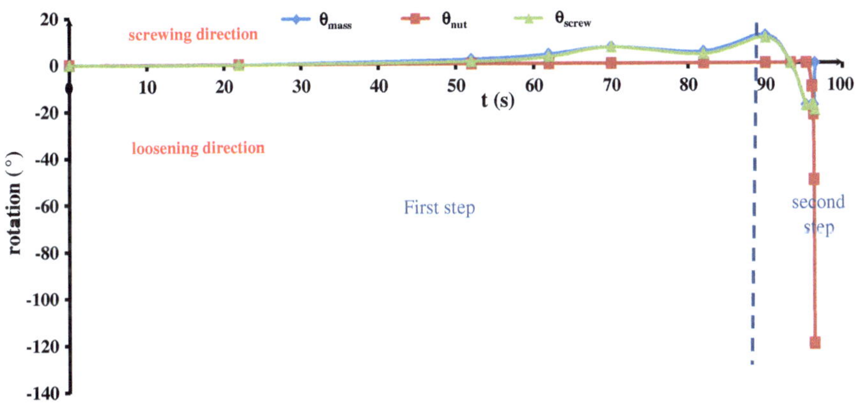

Fig. 6 Nut and screw rotation

In order to show the influence of the friction coefficient, a number of tests were elaborated for two types of screws (shown in Fig. 4). The variation of the axial load for the two cases is shown in Fig. 7. We can conclude that the loosening time for the galvanized screw was less than for than treated one. This can be explained by the friction coefficient, which was smaller for the galvanized screw. All experiments show that self-loosening happens always with galvanized screws. So the use of this type of screws in structures subject to dynamic effort or vibration increases the risk of loosening. This can be explained by the low friction coefficient associated with the galvanizing surface treatment, which reduces the risk of corrosion at the expense of the reliability of the assembly.

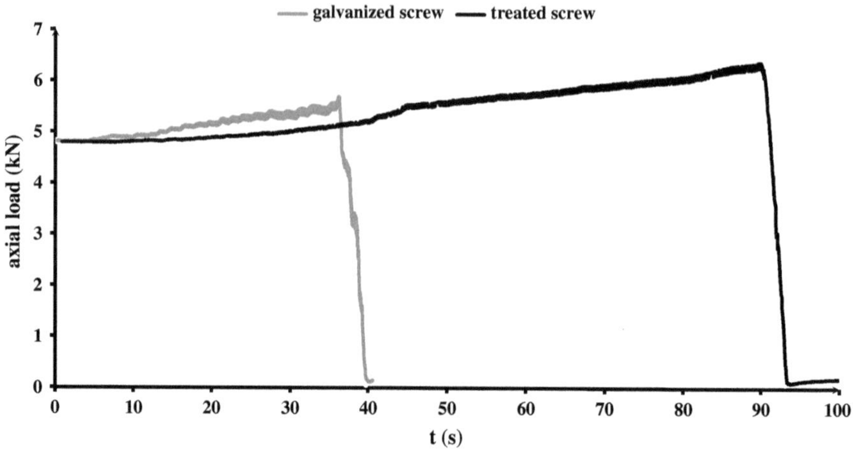

Fig. 7 Axial load variation for two types of screw

4 Conclusion

Bolt joints are used in many structures, and self-loosening sometimes occurs when they are subjected to transverse load. Most experiments are based on the Junker machine, which imposes relative displacements between the assembled parts. In this paper, an experimental approach was used to reveal the bolted joint self-loosening under vibrations. A transverse reciprocating acceleration was applied to an inertial mass, which initiated sliding under the bolt. Results show the progress of rotation of the different parts, with unscrewing of the bolt and loss of tension in the assembly. This provides a way to explore the loosening process of various types of bolts, under realistic conditions of vibration. We showed that galvanized screws loosen more easily than uncoated ones.

In our experimental device, the size and initial tightening of the bolts under test was limited by the force available from the shaker. We plan further experiments on a new device allowing higher loads to be used on larger bolts. The results will be compared with FEM simulation of dynamic behaviour in order to set and validate the self-loosening models, in the aim of obtaining a predictive model of self-loosening assemblies.

References

Bhattacharya A, Sen A, Das S (2010) An investigation on the anti-loosening characteristics of threaded fasteners under vibratory conditions. Mech Mach Theory 45:1215–1225. doi:10.1016/j.mechmachtheory.2008.08.004

Aziz H (2003) Etude du dévissage spontané des assemblages boulonnés. Ph.D. thesis, Institut National des Sciences Appliquées de Toulouse

Bouchaala N, Dion JL, Peyret N, Haddar M (2013) Micro-slip induced damping in the contact of nominally flat surfaces. Int J Appl Mech 5:1–20. doi:10.1142/S1758825113500051

Dinger G, Friedrich C (2011) Avoiding self-loosening failure of bolted joints with numerical assessment of local contact state. Eng Fail Anal 18:2188–2200. doi:10.1016/j.engfailanal.07.012

Dong Y, Hess DP (2000) Shock induced loosening of dimensionally non-conforming threaded fasteners. J S V 231:351–359

Goodier JN, Sweeney RJ (1945) Loosening by vibration of threaded fastenings. Mech Eng 67:798–802

Hess DP, Sudhirkashyap SV (1997) Dynamic loosening and tightening of a single bolt assembly. J Vib Acoust 119:311–316

Hess DP (1998) Vibration and shock induced loosening. In: Bickford JH, Nasser S (eds) Handbook of bolts and bolted joints. Marcel Dekker, New York, pp 757–824

Junker GH (1969) New criteria for self-loosening of fasteners under vibration. SAE Trans 78:314–335

Nassar SA, Yang X (2009) A mathematical model for vibration-induced loosening of preloaded threaded fasteners. J Vib Acoust 131:1–13. doi:10.1115/1.2981165

Granular Material for Vibration Suppression

Marwa Masmoudi, Stéphane Job, Mohamed Slim Abbes and Imad Tawfiq

Abstract We present preliminary results devoted to study and analyze the main features of particle damping from numerical simulations and experimental procedures. Particle damping is measured for free-free beam with particle filled enclosure attached to their both sides. Particle damping is measured for different amplitude vibration and the goal is to highlight the damping phenomena and to understand such mechanism involved in particle damping.

Keywords Particle damping · Discrete element method · Experimental validation · Vibration reduction

1 Introduction

Particle damping is a passive vibration damping, constituting an emerging technology for the control and the attenuation of mechanical vibration.

Particle damping provides vibration suppression by inserting non-cohesive particles in an enclosure attached or embedded in a vibrating structure, as illustrated schematically in Fig. 1. When the structure vibrates, the grains moves freely and collides between each other and with the enclosure walls. All these inherently inelastic and frictional collisions lead to energy dissipation. This mechanism differs from traditional damping, in which the strain energy stored in the structure is generally converted into heat by means of viscoelastic damper, such as elastomers; here, frictional and plastic collisions mechanisms are weakly dependent on

M. Masmoudi (✉) · M. S. Abbes
LA2MP, Enis, Route Soukra Km 3.5, BP 1173, 3038 Sfax, Tunisie
e-mail: masmoudi.marwa@gmail.com

S. Job · I. Tawfiq
LISMMA EA 2336, Supméca, 3 rue Fernand Hainaut, 93407 Saint-Ouen, France

M. S. Abbes et al. (eds.), *Mechatronic Systems: Theory and Applications*, Lecture Notes in Mechanical Engineering, DOI: 10.1007/978-3-319-07170-1_5,

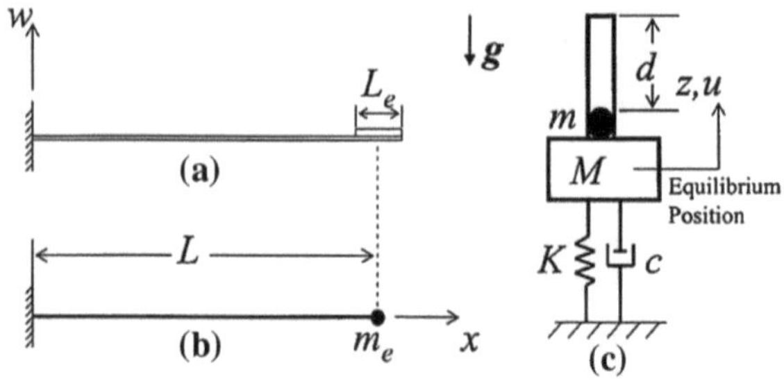

Fig. 1 A model of a beam with an attached enclosure filled with grains

the frequency and highly dependent on the excitation level: both occurs in addition to viscoelastic properties of the particles' bulk material. Particle dampers thus constitutes very powerful dissipators, over a broad frequency ranges. The particle damping technology has been used in many fields for vibration reduction such as the aerospace industry, the automotive and oil and gas industries.

Particle damping is a derivative of single-mass impact damper that have been studied by Popplewell and Semergicil (1989), Papalou and Masri (1998). Particle dampers are now more adopted than impact dampers due to the lower noise they produce and the lower wear of contact surfaces.

It is an attractive alternative in passive damping due to its conceptual simplicity, potential effectiveness over broad frequency range, temperature and degradation insensitivity, and cost effectiveness. These characteristic have been developed by many researchers such Panossian (1992), Friend and Kinra (2000), Simonian (1995).

There are significant numbers of design parameters that can affect the performance of the particle damper such as the particle size, shape, number, and density, in addition to the size and shape of the enclosure, and the amplitude excitation and frequency. Most of these parameters have been studied by many authors like Friend and Kinra (2000), Saeki (2002).

2 Modeling Techniques

There are several different ways in which researchers have attempted to model the behavior of particle dampers.

2.1 Analytical Methods

Many authors used a simplified model to describe the behavior of particle damping. They assumed that all particles move as a lumped mass, i.e. modeling a particle damper as an impact damper. This approach is used by Friend and Kinra

(2000) who idealized the beam as a standard Euler–Bernoulli beam vibrating in its fundamental mode, and the enclosure as a point mass attached to the tip of the beam. Then, the continuous beam is reduced to an equivalent SDOF, as shown in Fig. 1

The reduced mass of the equivalent SDOF system is expressed as

$$M = 0.24\rho AL + m_{encl} \tag{1}$$

Similarly, the reduced stiffness of the vibrating beam is defined by

$$K = \frac{3EI}{L^3} \tag{2}$$

And the reduced damping coefficient of the beam is given by

$$C = \frac{\psi_b}{2\pi}\sqrt{KM} \tag{3}$$

With Ψ_b is the intrinsic material damping of the beam material.

Specific damping capacity (C) is defined as the kinetic energy converted into heat during one cycle (ΔT) relative to the maximum kinetic energy of the structure during the cycle (T),

$$C = \frac{\Delta T}{T} \tag{4}$$

where ΔT is the energy dissipated during an impact and is expressed as

$$\Delta T = \frac{1}{2}(1 - R^2)\frac{m}{1+\mu}(v_p^- - v_2^-)^2 \tag{5}$$

With μ is the ratio of the mass of the particle to the primary system and R is the coefficient of restitution.

The maximum kinetic energy of the structure is defined as

$$T = \frac{1}{2}\left(\frac{M_r m}{M_r + m}\right)(v_p^- - v_2^-)^2 \tag{6}$$

2.2 *Particle Dynamics Methods*

In the following, a summary of the theory developed by Olson (2003) is presented.

In his chapter, Olson uses the particle dynamics method to model the behavior of particle dampers and to analyze the loss mechanisms involved in particle damping.

The particle dynamics method is a method in which each particle is modeled and their motion is tracked. The procedure is an explicit process with an appropriate time step. During a single time step, disturbances propagate between neighbors particles. At any time, the resultant forces on any particle are calculated by its interaction with the particles with which it is in contact. Interaction forces between particle-particle and particle-enclosure walls are calculated based on prescribed force–displacement relations. For accurate damping predictions, one must incorporate dissipation into the model. For instance, the energy is dissipated owing to the viscoelastic behavior of the sphere material. Here, the dissipative portion of the normal force would be precised in terms of a viscoelastic formulation of Hertz's theory or in terms of Hertz-Mindlin (Di Renzo and Di Maio 2005) potential if one accounts for tangential friction. The particle damper model has been implemented within X3D.

2.3 DEM Methods

Many authors have used the discrete element method (DEM) to describe systems of particle damping and to study the damping mechanisms of these devices in vibration problem. In DEM modeling, the trajectory of each granule is tracked incrementally by Newton's equation of motion. Forces are computed at the contacts by using an appropriate contact model for inter-particle and particle–cavity impact and friction interaction. Using a linear model that uses a combination of a spring and a dashpot is a simplest model, though being fairly inaccurate to predict the correct forces, this approach has been pioneered by Cundall and Strack (1979) and used by Cundall and Strack (1979), Cleary (2000). Various non-linear model developed in the literature (Cleary 2000; Mishra and Murty 2001; Zhang and Vu-Quoc 2002; Johnson 1985) can be used for better precision in describing the nonlinear impact phenomenon. A nonlinear contact model is described below and summarized in Fig. 2. The normal component on the contact force can be modeled by the sum of a spring using the theory Hertz and a damping force. The tangential force is composed by a combination of a spring and a dashpot if sticking and can be described by coulomb's friction if sliding.

$$F_{ij}^{n} = k_n \delta_{nij}^{\frac{3}{2}} + c_n \dot{\delta}_n \tag{7}$$

$$F_{ij}^{t} = k_s \delta_{sij}^{\frac{3}{2}} + c_s \dot{\delta}_s \ (if\ sticking) = \mu f_n \ (if\ sliding) \tag{8}$$

As the structure vibrate causing the collision and friction between granules themselves and between granule and the enclosure wall, the granule may have the translational and the rotational motion and are expressed by

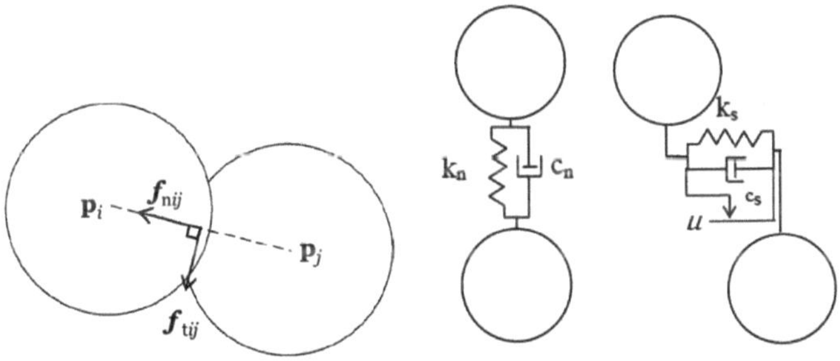

Fig. 2 The contact models between particles in normal and shear directions

$$m_i \ddot{p}_i = m_i g + \sum_{j=1}^{k} (F_{ij}^n + F_{ij}^t) \tag{9}$$

$$I_i \ddot{\theta}_i = \sum_{j=1}^{ki} T_{ij} \tag{10}$$

where m_i the mass of particle i, I_i is the moment of inertia of particle i and g is the gravity acceleration, p_i is the position vector of the center of gravity of the granules, θ_i is the angular displacement vector, F_{ij}^n is the normal contact force between particles i and j F_{ij}^t is the tangential contact force, k_i the total number of contact with particle i.

3 Experimental Procedure

We report here our very first attempt to probe the capacity of particle dampers inserted in a deformable beam. The beam is made of Aluminum and its dimensions are: length 1000 mm, width 4 cm and height 0.3 cm. On both sides of the beam, four particle containers are placed equidistantly. The whole device is excited with a shaker connected to voltage generator via a power amplifier. The signal used to excite the structure is a 3 s sweep sine between 5 and 500 Hz.

To measure the acceleration, an accelerometer of type Sigma 256HX-10 is placed on top of the beam, in its center. The system is also equipped with a force sensor of type Kistler 9011A placed in between the shaker and the beam, to probe the excitation force. The material of the particles is glass beads and their diameters are about 1 mm. The thickness of the particle bed is 5 mm. The containers have a cubic form and are made with paper sheet; their dimensions are $10 \times 10\ \text{mm}^2$.

A picture of the experimental setup is shown in Fig. 3.

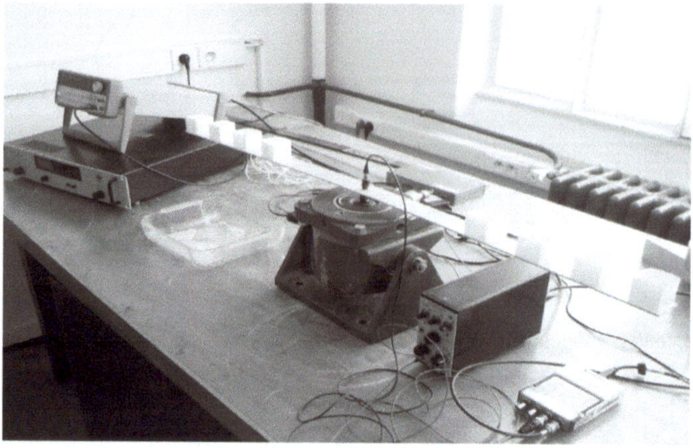

Fig. 3 Picture of the experimental setup, showing a beam on a shaker and eight particle dampers cells

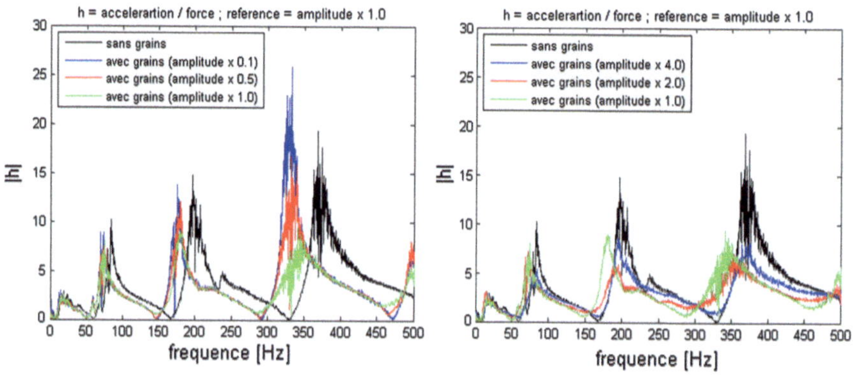

Fig. 4 Comparison of the response beam with and without particles for different level excitation

4 Experimental Results

Our experimental setup allows us to estimate the transfer function between the force exerted on the beam and its acceleration, and compare the response of a beam with and without particles, as illustrated in Fig. 4.

At first glance, the results shown in Fig. 4 demonstrate that the particle dampers cause a significant decrease in the amplitude response of the structure.

More precisely, we ran a series of measurements by varying the amplitude of the excitation. When we increased the level of excitation the particle dampers were significantly more effective in lowering the peaks of the transfer function; this reflects that particles have tendency to collide more with each other, which increases the energy dissipation.

By decreasing the reference amplitude one can see only the effect of an added mass: the grains do not moves at all in their enclosures.

In turn, the effects of adding the particle dampers to the beam lowers its natural frequency but under high level excitation the granular bed shows a lower effective mass and higher resonance frequency, since the particles are in motion and do not stay in contact with the beam.

5 Conclusion

We first exposed the main features required to simulate particle dampers and we underlined the basic mechanisms under play. We then presented preliminary experiments devoted to demonstrate the effectiveness and the relevancy of particle dampers to control the vibrations of a simple beam. This work is our first attempt to better understand these systems.

References

Cleary PW (2000) DEM simulation of industrial particle flows: case studies of dragline ex cavators, mixing in tumblers and centrifugal mills. Powder Technol 109:83–104

Cundall PA, Strack ODL (1979) The distinct numerical model for granular assemblies. Geotechnique 29:47–65

Di Renzo A, Di Maio FP (2005) An improved integral non-linear model for the contact of particles in distinct element simulations. Chem Eng Sci 60:1303–1312

Friend RD, Kinra VK (2000) Particle impact damping. J Sound Vib 233:93–118. http://dx.doi.org/10.1006/jsvi.1999.2795

Johnson KL (1985) Contact mechanics. Cambridge University Press, Cambridge

Mishra BK, Murty CVR (2001) On the determination of contact parameters for realistic DEM simulations of ball mills. Powder Technol 115:290–297

Olson SE (2003) An analytical particle damping model. J Sound Vib 264:1155–1166

Panossian HV (1992) Structural damping enhancement via non-obstructive particle damping technique. ASME J Vib Acoust 114:101–105

Papalou A, Masri SF (1998) An experimental investigation of particle damper under harmonic excitation. J Vib Control 4:361–379

Popplewell N, Semergicil SE (1989) Performance of bean bag impact damper for a sinusoidal external force. J Sound Vib 133(2):193–223

Saeki M (2002) Impact damping withgranular materials in a horizontally vibrating system. J Sound Vib 251:153–161. http://dx.doi.org/10.1006/jsvi.2001.3985

Simonian SS (1995) Particle beam damper. In: Proceedings of SPIE conference on passive damping, SPIE, vol 2445, SPIE, San Diego, CA, pp 149–160

Zhang X, Vu-Quoc L (2002) Modeling the dependence of the coefficient of restitution on the the impact velocity in elasto-plastic collisions. Int J Impact Eng 27:317–341

Influence of Trust Evolution on Cost Structure Within Horizontal Collaborative Networks

Omar Ayadi, Garikoitz Madinabeitia, Naoufel Cheikhrouhou and Faouzi Masmoudi

Abstract Partner trust is one of the most critical factors influencing the performances of collaborative networks guaranteeing the success of the relationships between the enterprises of a network. In this chapter, based on a classification of trust types with respect to trust influencing factors, a rule-based fuzzy system assessing the influence of each trust type on collaborative networks related costs is designed. Results show that each considered cost evolves not only differently with respect to the evolution of specific trust types but also independently from each other. Moreover, the developed system provides quantitative predictions of costs reductions that can be reached by collaborative networks partners based on their corresponding trust levels.

Keywords Trust · Cost reduction · Fuzzy logic · Collaborative networks

O. Ayadi (✉) · F. Masmoudi
Laboratoire de Mécanique, Modélisation et Production, Ecole Nationale d'Ingénieurs de Sfax, Université de Sfax, Route de Sokra km 3.5, 3038 Sfax, Tunisia
e-mail: omar.ayadi@yahoo.fr

F. Masmoudi
e-mail: faouzi.masmoudi@enis.rnu.tn

G. Madinabeitia · N. Cheikhrouhou
Laboratory for production management and processes, Ecole Polytechnique Fédérale de Lausanne, Station 9, 1015, Lausanne, Switzerland
e-mail: garthymad@hotmail.com

N. Cheikhrouhou
e-mail: naoufel.cheikhrouhou@epfl.ch

M. S. Abbes et al. (eds.), *Mechatronic Systems: Theory and Applications*, Lecture Notes in Mechanical Engineering, DOI: 10.1007/978-3-319-07170-1_6,

1 Introduction

In view of the evolution of dynamic markets and the increasing competitiveness, companies are more willing to cooperate within specific organizational frameworks such as virtual organizations, virtual breeding environments and collaborative networks. Particularly, horizontal collaborative networks are associations of independent organizations, with similar core competences, called members that come together and share resources and skills to achieve common goals/missions, such as acquiring and executing business opportunities or reducing costs (Barratt 2004). These types of cooperation allow companies to take benefits from the competences of each other and to act collaboratively in order to reduce costs and to stay competitive. However, these favours necessitate the satisfaction of certain levels of trust between collaborating partners. In fact, the establishment of trust has been identified as the most critical factor that facilitates partnering success in collaborative networks (CNs) (Cheikhrouhou et al. 2012a). Increasing trust levels between partners provides possibilities to improve different aspects of a relationship. In order to master the evolution of the CN, partners need to estimate the cost reduction that can be reached from their collaboration with respect to the current trust level in each other.

The objective of this chapter is to design and develop a rule-based fuzzy logic (RBFL) system evaluating the influence of the evolution of the trust attributes on cost reduction within horizontal CN. This chapter first presents a classification of trust types based on a categorization of trust influencing factors. This classification is then considered in the design of the RBFL system evaluating the influence of each trust type on CN related costs.

The chapter is outlined as follows: the Sect. 2 presents a state of the art concerning trust types and the relationship between trust and the economic performances. In Sect. 3, the proposed methodology is described and developed. A classification of trust types is presented in Sect. 4, and the designed RBFL system is addressed in Sect. 5. Section 6 reports the results and corresponding discussions. Finally, the chapter ends with conclusions and future research directions.

2 State of the Art

Different aspects or attributes of trust have been considered in the literature, highlighting the important relationship between trust and the competitiveness/ performances of collaborative organizations. Since trust is a behavioural issue, trust concepts, as considered within collaborative organizations, are usually linked to the application field and/or to the partner location. Different examples can be provided such as the influence of trust in construction projects (Wong and Cheung 2004; Kadefors 2004), the relationship between small sized businesses and service providers in health insurance, management consulting or telecommunication

(Coulter and Coulter 2003), motor industry in UK (Leverick and Cooper 1998), logistics problems in China (Tian et al. 2008), maintenance services for paper industry in Finland (Laaksonen et al. 2009), measure of trust in virtual project teams (Chen et al. 2008; Hung et al. 2004), the identification of trust attributes and antecedents for automaker suppliers in Korea, Japan and USA (Dyer and Chu 2000) and Australian meat and horticulture supply chain (Paterson et al. 2008).

Several trust definitions and different representative types of trust can be found in the literature using different perspectives. Trust can be seen as an one-dimensional construct as considered by (Schmidt et al. 2007) who define it as "the requesting agent's belief in the recommendation queried agent's willingness and capability to behave as expected by the requesting agent in a given context at a given time slot". From the same perspective, another definition is considered by (Bharadwaj and Al-Shamri 2009) and (Ayadi et al. 2010, 2013) that consider trust as the extent to which one party is willing to participate in a given action with a given partner, considering the risks and incentives involved. From the perspective that considers trust as a multifaceted concept, several trust types are proposed in the literature. A review of the literature permits to exhaustively identify the trust types and their basic definitions as follows:

- Competence/expertise trust: the perception of others to perform the required tasks based on observable proofs, experiences or connections with professional bodies (Hartman 2003).
- Contractual trust: the expected degree of honesty to respect and honour the agreement (written or oral) (Paterson et al. 2008).
- Relational/personal trust: the human aspects of an existing relationship between human resources of collaborating institutional partners (Coulter and Coulter 2003; Kadefors 2004).
- Cognitive trust: the degree of partner's integrity and ability within a relationship (Greenberg et al. 2007).
- Intuitive trust: the emotional aspect of trust founded upon the party's prejudice, biases or other personal feelings towards the counter parts (Hartman 2003).

In order to evaluate trust in virtual teams (Chen et al. 2008) classify trust into three categories:

- Direct trust is derived from the direct relationship and interactions existing between two virtual team members.
- Indirect trust represents the trust component between two virtual team members that can be calculated through the direct trust existing between them and a third member. This third member can be in interaction with one of the two considered members or with both of them.
- Negative trust is defined as the level of untrustworthiness of one virtual team member for another. It reflects the relationship aspects that can negatively influence the trust level between virtual team members.

Research papers dealing with trust benefits show that the trust level between cooperating or collaborating partners can influence the economic performances of the relationship existing between the partners (Barratt 2004). (Laaksonen et al. 2009) develop a game theoretic approach in order to model the cost structures of three supplier-buyer relationships. This study shows that mutual high trust leads to high financial benefits, in contrast to the zero utility of mutual low trust. (Zaghloul and Hartman 2002) conduct a survey across the Canadian construction industry showing that the improvement of trust between contracting parties in this field can yield projects' total cost savings. Based on a hypothesized model validated and statistically analyzed through a survey conducted across 980 UK manufacturers of electronic equipments (Panayides and Venus Lun 2009) show that trust can significantly improve supply chain performances and reduce costs. (Huang and Li 2001) propose three advertising models using a game theoretic approach. Their findings show that retailers utilize co-operative advertising to reduce their total promotional expense substantially by sharing the cost of advertising with the manufacturer, demonstrating that the system profit is higher if cooperation exists. Based on the literature investigation and analyse and on interviews conducted with top managers of a multinational corporation (Beccerra and Gupta 1999) show that high trust levels can reduce transaction and agency costs in organizations. (Fawcett et al. 2006) report the statment of supply practitioners affirming that increasing trust allow the reduction of transaction costs. (Tian et al. 2008) propose a conceptual model validated through interviews with academic and industrial experts showing that the increase of trust level in China's third-party logistics improve the efficiency of these practitioners by reducing the substantial monitoring costs. (Chiles and McMackin 1996) analyze the literature and provide some theoretical propositions dealing with how trust can decrease transaction costs. Nevertheless, these propositions need to be empirically validated. (Zaheer et al. 1998) propose a hypothesised model tested with data gathered from 107 supplier-buyer inter-firm relationships in the electrical equipment manufacturing industry. Their findings show that both inter-personal and inter-organizational trust influence negotiation costs whose reduction is more dependent on inter-organizational trust than on inter-personal trust.

The literature shows the existence of strong and highly complex relationships between trust and economic performances, in particular with cost reduction within cooperating and collaborating structures. Nevertheless, there is a lack in understanding the cost structure in CNs and its evolution with respect to the evolution of trust level. A detailed study based on a classification of trust types is necessary in order to identify the influence of each trust type on specific costs in CNs. Moreover, very few researchers address the quantitative impact of trust on cost or benefit in networked organizations; in particular the dynamics of trust (evolution) is ignored. In this chapter, a RBFL system is developed to model the relationships between trust types and the economic performances in CNs. The developed system is a predictive model that quantitatively estimates the influence of trust types' evolution on specific CN costs, namely purchasing cost, quality control cost and marketing cost. Based on the classification of trust types, the system predicts the influence of each trust type on the considered CN costs and estimates the cost reduction that can be reached with respect to the evolution of trust types' values.

The experience of the Swiss CN "Swiss Microtech" (SMT) is considered in the development of the RBFL system. In fact, SMT has made considerable profits from trust levels characterizing the relationships between its partners.

3 Proposed Methodology

The objective of this chapter consists of designing a tool predicting the influence of trust and its evolution on CN costs namely purchasing cost, quality control cost and marketing cost. Based on the literature review, this chapter first define an exhaustive list of factors influencing trust considered as a basis for the design of a classification of trust types based on its influencing factors. The defined list of factors is considered in order to identify the trust types influencing the considered costs. Next, interviews are conducted with experts in order to identify the relationship between trust types and costs reductions. With respect to the results of the interviews, a RBFL system is designed, linking the classified trust types to the corresponding costs, in order to assess the influence of each trust type on the evolution of the considered costs. The rules managing the RBFL system are developed with the contribution of industrial experts from SMT.

4 Classification of Trust Attributes with Respect to Their Types

(Cheikhrouhou et al. 2012b) propose a classification of the different trust attributes with respect to the representative trust categories. The following attributes are taken into account in the classification: product/service quality, timeliness, reliability, spirit of cooperation, product/service customization, transparency, confidentiality, honesty, information sharing, shared value, commitment to the relationship, benevolence, predictable behaviour, friendliness, reputation, work standards, financial stability, employees qualification, partnership duration, asymmetry of the relation, opportunistic behaviours and own specific asset. For more details about the cited factors, it is advised to look for the research works by: (Wong and Cheung 2004; Tian et al. 2008; Kwon and Suh 2004; Coulter and Coulter 2003; Mun et al. 2009; Paterson et al. 2008). The classification of trust types based on the influencing factors provides 5 groups: competence trust, contractual trust, relational trust, indirect trust and negative trust. Table 1 summarizes this classification and links each type to its influencing factors.

Competence trust "T_{cp}" describes the expected degree of partner's competence in terms of product/service quality and the respect of delivery time. This trust type expresses the most general appreciation of the work of the partner. The main factors related to it are: product/service quality, timeliness/punctuality (respect of

Table 1 Classification of trust types (Cheikhrouhou et al. 2012b)

Trust types	Influencing factors
Competence trust	Quality, timeliness, reliability
Contractual trust	Spirit of cooperation, customization, transparency, confidentiality, honesty
Relational trust	Information sharing, shared value, commitment to the partnership, benevolence and supportiveness, predictable behavior, friendliness and politeness
Indirect trust	Reputation, work standards, financial stability, qualification of employees, duration of partnership
Negative trust	Dependence and asymmetric relation, opportunistic behavior, own specific asset

deadlines) and reliability (the ability to perform the required tasks under stated conditions during a specified time period).

Contractual trust "T_{ct}" expresses the expected degree of honesty to fulfill and honor the agreements, and takes into account all agreement types regardless of their forms, whether or not they are written in contracts. The influencing factors are: cooperation spirit, customization of products or services with slight variations of standard configurations that are typically developed in response to a specific order by a customer, transparency, confidentiality and honesty.

Relational trust "T_r" reveals from the existing relationships between people working in cooperating enterprises and considers mostly the human aspects of the relationship. It can be influenced by the following factors: information sharing representing the willingness to share information for the good of the partnership, shared value, commitment to the relationship, benevolence and supportiveness, predictable behavior, and friendliness and politeness.

The indirect trust "T_i" is the partner confidence that can be reached due to the partner's history and actions with external parties. It regroups the external factors of the partnership that can be summarized in the following attributes: reputation, work standards, financial stability, qualification of employees, and duration of partnership.

Negative trust "T_n" is the level of untrustworthiness between two partners. The negative trust considers the relationship aspects that can influence negatively the partnership. This trust type can be related to a difference of power between the partners in a relation when the powerful partner can take advantage of his power. More the relation is unbalanced, higher the negative trust will be. This power asymmetry can lead to behaviors that are harmful for a relation and can sap it. The principal factors of negative trust are the asymmetry of a relation, opportunistic behaviors and own specific asset. Asset specificity refers to investments in physical or human assets that are dedicated to a particular business partner and whose redeployment entails considerable switching costs (Kwon and Suh 2004).

The trust types discussed in (Cheikhrouhou et al. 2012b) do not include explicitly three trust types: the cognitive trust, the intuitive trust and the direct trust. This does not mean that these types are neglected, but they are implicitly considered in order to omit the redundancy of trust aspects in the classification. Indeed, cognitive trust expresses the partner integrity and ability within the relationship, when these two factors are affected to the competence trust (through the reliability factor) and the relational trust (through shared value and commitment to the relationship factors). Furthermore, intuitive trust can be considered as a part of relational trust that is represented by predictable behavior and friendliness and politeness factors. Finally, direct trust as defined in the literature expresses the trust component that is related to the direct relationship between the collaborating partners. So, this type encloses all the following considered types: competence trust, contractual trust and relational trust. In this chapter, these types are considered separately in order to better evaluate their influence on cost evolution.

Moreover, the analysis of the classification of trust types shows that the increase (resp. the decrease) of negative trust will be expressed by the decrease (resp. the increase) of contractual and relational trust levels. In fact, negative trust is affected by three factors (the asymmetry of the relation, opportunistic behavior and own specific asset) those evolutions are directly linked to the evolution of contractual and relational trust factors: the asymmetry of the relation is related to transparency, information sharing and commitment; opportunistic behavior are linked to benevolence and predictable behavior; and own specific asset is in direct relation with shared value and commitment, that are factors of contractual and relational trust. The overlap of these interrelated trust types in the RBFL system can lead to noisy effects on the results. For this reason, the level of negative trust will not be explicitly considered in the classification of the trust types and in the RBFL system inputs. Its influence on costs reduction will be shown through the influence of contractual and relational trust levels. As a consequence, the trust types' classification that is considered in this chapter contains four types: competence trust, contractual trust, relational trust and indirect trust.

5 Development of Rule-Based Fuzzy System Assessing the Influence of Trust Types on Economic Performances

The most important economic performances addressed by a horizontal CN are purchasing cost, quality control cost and marketing cost. Purchasing cost represents the cost of raw materials that partners need to purchase from external companies. Quality control cost includes costs of controlling materials transferred between two partners of the CN or more. Marketing cost includes all fees of marketing activities that partners carry out such as advertisements, participation to expositions and flyers design, creation and diffusion. Accordingly, the RBFL system is designed and developed in order to assess the influence of the trust types on these costs. The system inputs are the levels of trust between collaborating partners and the expected outputs are the percentages of cost reduction that could be reached by the network. Structured and unstructured interviews are conducted

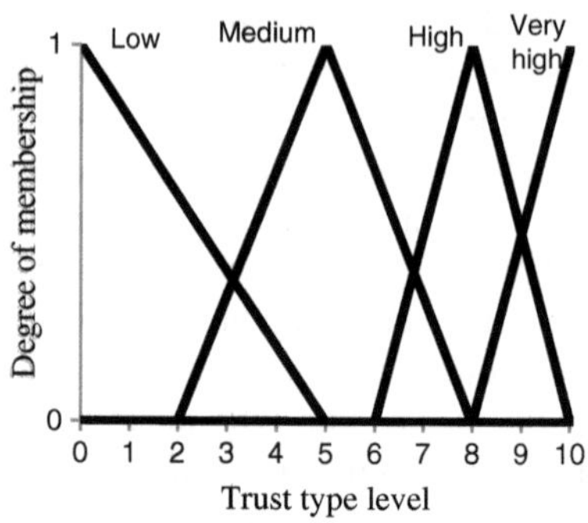

Fig. 1 Membership function of each trust type

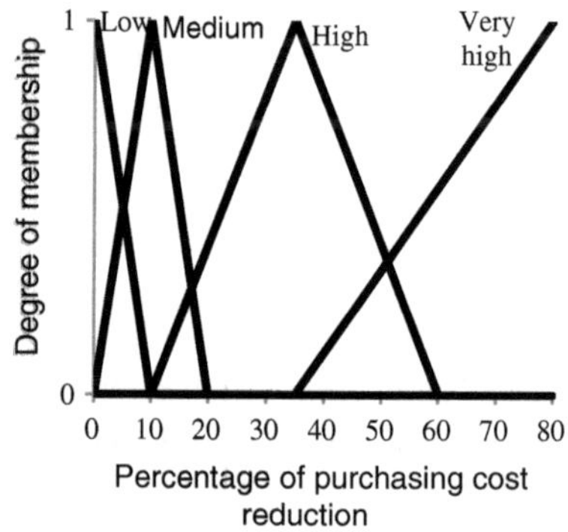

Fig. 2 Membership function of purchasing cost reduction

with both industrial and academic experts who have gained important experience with the considered CN "SMT" since its creation. These interviews allow us to well understand the relationships between the trust types and the considered costs from an empirical point of view. Each cost category is then linked to its influencing trust types using fuzzy rules. The rules are checked and validated by these experts.

The system inputs are the levels of trust types, namely competence trust, contractual trust, relational trust and indirect trust. The trust level for each type is evaluated by the CEOs of the companies taking part to the CN. The evaluation is a subjective assessment based on their judgments. In addition, the evolution of trust types' levels does not directly affect the economic performances of CN, but it enables the decision makers to take decisions influencing the considered costs, such as sharing information about the cheaper supplier or pooling the quality control operations. As a consequence, fuzzy logic is used as an appropriate technique tackling the vagueness and the impreciseness of human evaluations. Each trust type is represented by a triangular membership function (TMF) with four linguistic variables: (low, medium, high, very high) as defined in Fig. 1.

The percentages of each cost reduction as output of the RBFL system is also represented by TMFs, with four triangular fuzzy numbers (low, medium, high, very high) for purchasing cost reduction (Fig. 2) and quality control cost reduction (Fig. 3), and with three triangular fuzzy numbers (low, medium, high) for

Fig. 3 Membership function of quality control cost reduction

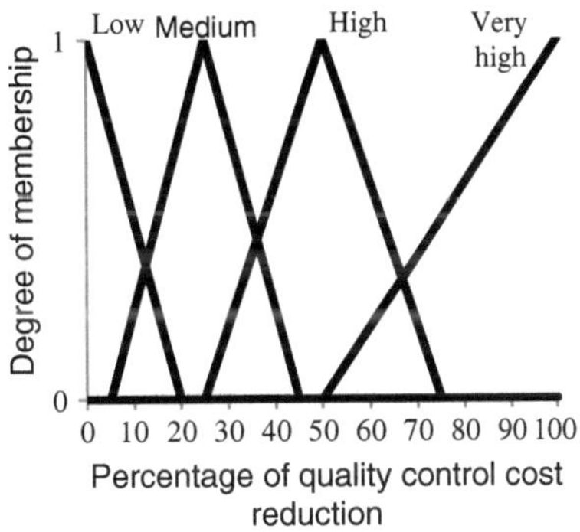

Fig. 4 Membership function of marketing cost reduction

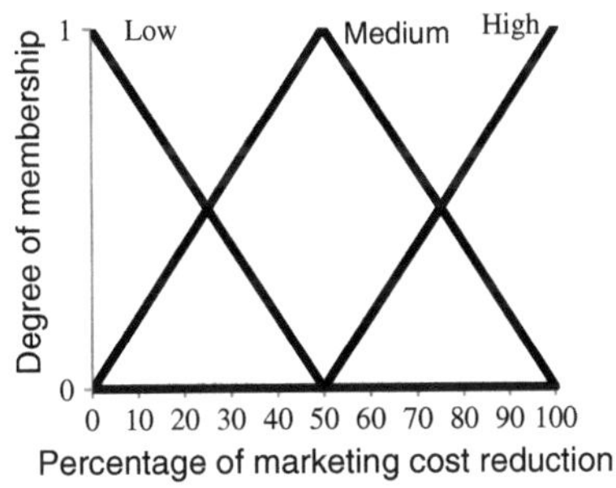

marketing cost reduction (Fig. 4). As shown in Figs. 2, 3 and 4, the membership functions are characterized by a maximum purchasing cost reduction equal to 80 %, whereas those of quality control and marketing costs can reach 100 %. This can be explained by the inexistence of any action that can lead to fully eliminate purchasing cost, whatever is the trust level between collaborating partners. However, quality control costs (respectively marketing costs) can be decreased to 0 if the partners decide to not proceed to quality control of materials transferred between the corresponding factories (respectively to eliminate marketing activities if the companies consider that they have sufficient number of customers and there is no need to look after for clients).

The RBFL system uses IF-THEN rules formulated on the basis of the interview results that link the reduction of each considered cost to its influencing trust types. The opinions of interviewed experts converge to the corresponding statements:

- Purchasing cost depends only on relational trust
- Quality control cost depends on competence and contractual trust
- Marketing cost depends on relational, competence and indirect trust

86 rules are designed in order to cover all possible combinations of trust types' levels and define the corresponding costs reduction. Each rule defines the linguistic variable of the reduction of only one cost type: purchasing costs, quality control costs or marketing costs. The rules are weighted with a coefficient comprised between 0 and 1 with respect to their judged contribution in the designed system.

The Mamdani's fuzzy inference system (FIS), in (Mamdani and Assilian 1975), is adopted in this research work. In fact, it is well suited to human judgments and can be directly applied to logical rules made by human beings, rendering the leverage of existing expert knowledge easy (Chen et al. 2010; Sivanandam et al. 2007). The adopted FIS is based on four main steps:

1. Fuzzification: the corresponding semantic fuzziness and membership degree of the considered input variables are defined. This definition is based on the TMFs.
2. Rule evaluation: the fuzzified inputs are evaluated in the rule antecedents in order to generate fuzzy output for each rule.
3. Aggregation of rules' outputs: the fuzzy outputs are aggregated using the "maximum" aggregation method generating a fuzzy aggregated output.
4. Defuzzification: the aggregated fuzzy output is defuzzified in a crisp output representing the percentage reduction in cost that can be reached with respect to the system inputs. The "centroid" method is adopted for defuzzification.

6 Discussion of Results

Several tests are performed in order to identify the influence of the evolution of the trust types' levels on costs reduction. Figure 5 shows the evolution of the percentage of purchasing cost reduction "C_p" with respect to its only influencing trust type, which is relational trust "T_r". The results show that T_r has not a significant effect on C_p unless the T_r level exceeds the value of 2 on a scale of 1–10. In fact, for these low levels of T_r, C_p does not exceed 4 %. This reduction in the percentages can be obtained by any enterprise of the network by a simple negotiation with his/her suppliers without relying on the collaboration with his/her partners. However, if the level of T_r exceeds the value of 8, collaborating with partners can produce a discount on purchasing cost more than 10 % and could reach a maximum value of 29 %. These discounts show the importance of T_r in enabling a strong negotiation between the network and the external suppliers. With T_r levels between 2 and 8, C_p can reach values comprised between 4 and 10 %. Furthermore, results show that C_p does not exceed 30 % even if T_r reaches its maximum level. This observation can be justified by the fact that C_p depends on others factors than trust, which should be fulfilled in order to achieve more percentage reduction in the purchasing cost, C_p. In this work, only the influence of trust types on costs is considered, with abstraction of any other factor such as the life spans of products and the evolution of the market that could affect C_p.

Figure 6 shows the evolution of the quality control cost reduction "C_{qc}" with respect to its influencing trust types, namely contractual trust "T_{ct}" and competence trust "T_{cp}". Results show that a minimum level of 2 on a scale of 1–10 should be fulfilled for both T_{ct} and T_{cp} in order to exceed 10 % of C_{qc}. The curve for C_{qc} is non-monotone for the range of contractual trust level 3–8 and competence trust level of 3–6. This area is characterized by 20 % of reduction in the quality control cost.

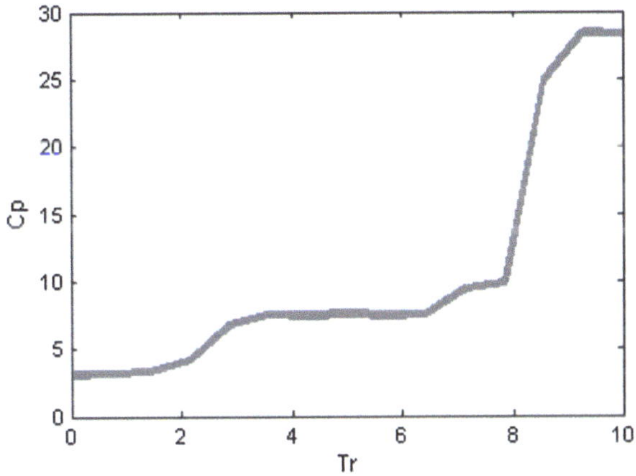

Fig. 5 Evolution of C_p with respect to T_r level

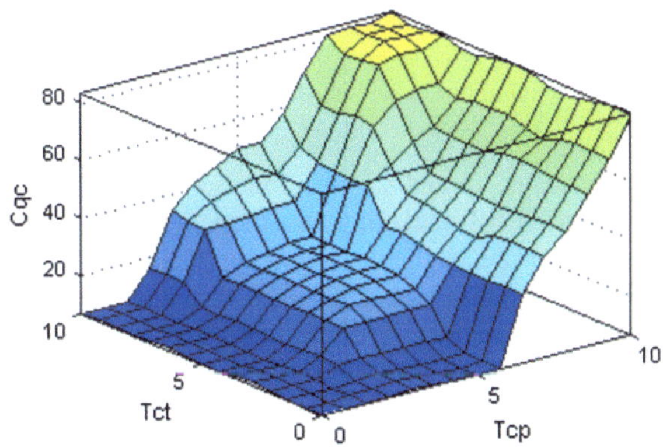

Fig. 6 Evolution of C_{qc} with respect to T_{ct} and T_{cp} levels

Moreover, results show that T_{cp} has more significant influence on C_{qc} than T_{ct}. In fact, even for a null T_{ct} level, C_{qc} can reach its maximum amount (80 %) if T_{cp} reaches its maximum level. However, C_{qc} does not exceed 50 % if T_{cp} level is less than 6, even for maximum level of T_{ct}. In order to benefit from important economic savings related to quality control cost, it is more interesting for CN decision makers to focus on improving T_{cp} by developing its influencing factors than on T_{ct}.

Figure 7 represents the evolution of marketing cost reduction "C_{mk}" with respect to T_{cp} and T_r levels for different levels of T_i. Figure 8 represents the evolution of C_{mk} with respect to T_i and T_r levels for different levels of T_{cp}.

Results show an equivalent dependence of C_{mk} on its influencing trust types. In the case one trust type is equal to zero (i.e. one of the types T_{cp}, T_r and T_i does not exist between collaborating partners), C_{mk} cannot exceed 10 % even if the two other trust types reach their maximum levels. This can be deduced from Fig. 7c ($T_i = 10$) for $T_r = 10$ and $T_{cp} = 0$ and for $T_r = 0$ and $T_{cp} = 10$, and Fig. 8c ($T_{cp} = 10$) for $T_r = 10$ and $T_i = 0$ and for $T_r = 0$ and $T_i = 10$, where C_{mk} is equal to 10 %.

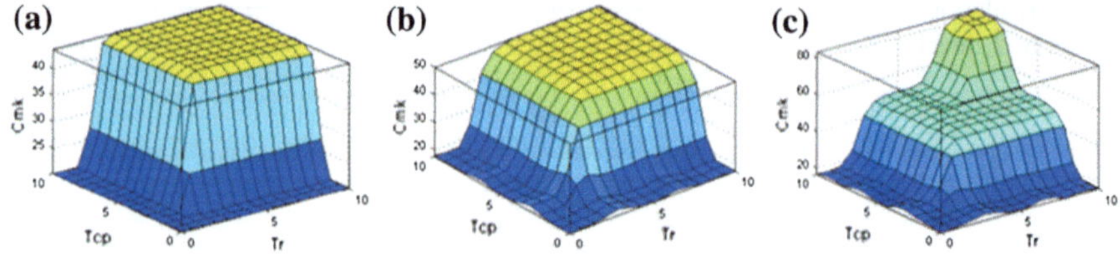

Fig. 7 Evolution of C_{mk} with respect to T_{cp} and T_r levels. **a** $T_i = 3$, **b** $T_i = 6$, **c** $T_i = 10$

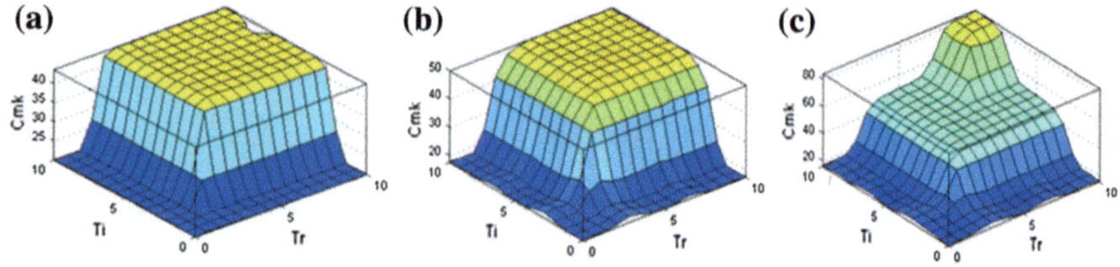

Fig. 8 Evolution of C_{mk} with respect to T_i and T_r levels. **a** $T_{CP} = 3$, **b** $T_{CP} = 6$, **c** $T_{CP} = 10$

From Figs. 7a and 8a, even low trust level (2–3 of T_{cp}, T_r and T_i) will results in at least 40 % of marketing cost reduction. This discount does not exceed 50 % if the level of one of the three trust types is less than 6 as shown in Figs. 7b and 8b. In addition, Figs. 7c and 8c show that reaching the maximum value of C_{mk} (i.e. 80 %) needs high trust levels: more than 8 for all influencing trust types.

7 Conclusions and Future Work

In this work, a rule based fuzzy system is developed in order to assess the influence of trust evolution on economic performances of CNs. The main scientific contribution of the developed system consists of quantifying the influence of each trust type on specific costs. The studied costs are considered as the most important economic performances addressed by a horizontal CN. This quantification takes into account the vagueness aspect of trust and its components, which is tackled by the use of fuzzy logic. The designed system represents a predictive model that estimates cost reduction percentages that can be reached by collaborating partners by the evolution of their trust in each others.

Moreover, based on this estimation, the developed system can be used in order to identify the most critical trust types to be monitored by the CN partners in order to reach interesting costs savings. In fact, results show that trust types influence differently the economic performances.

As future work, the designed system should be applied to real CNs. Addressing such implementation cases allow to check the feasibility of the proposed approach and to identify the eventual strategic characteristics of CNs that should be satisfied to reach predicted costs' reductions.

Furthermore, studying the evolution of each trust type with respect to its influencing factors represents another interesting research direction that allows a better understanding of the actions allowing trust level improvement and consequently, cost reduction.

References

Ayadi O, Cheikhrouhou N, Masmoudi F (2010) Assesment of customer trust level in supply chains based on information sharing dimensions. In: 8th international conference of modeling and simulation, Hammamet, Tunisia, 10–12 May 2010

Ayadi O, Cheikhrouhou N, Masmoudi F (2013) A decision support system assessing the trust level in supply chains based on information sharing dimensions. Comput Ind Eng 66(2):242–257

Barratt M (2004) Understanding the meaning of collaboration in the supply chain. Supply Chain Manage Int J 9(1):30–42

Beccerra M, Gupta AK (1999) Trust within the organization: integrating the trust literature with agency theory and transaction costs economics. Public Adm Q 23(2):177–203

Bharadwaj KK, Al-Shamri MYH (2009) Fuzzy computational models for trust and reputation systems. Electron Commer Res Appl 8(1):37–47

Cheikhrouhou N, Pouly M, Huber C, Beeler J (2012a) Lessons learned from the lifecycle management of Collaborative Enterprises Networks: the case of Swiss Microtech. Int J Manuf Technol Manage 23(8): 1129–1150

Cheikhrouhou N, Pouly M, Madinabeitia G (2012b) Trust categories and their impacts on information exchange processes in vertical collaborative networked organizations. Int J Comput Integr Manuf 25(12):1134–1147

Chen TY, Chen YM, Chu HC (2008) Developing a trust evaluation method between co-workers in virtual project team for enabling resource sharing and collaboration. Comput Ind 59:565–579

Chen TY, Chen YM, Lin CJ, Chen PY (2010) A fuzzy trust evaluation method for knowledge sharing in virtual enterprises. Comput Ind Eng 59(4):853–864

Chiles TH, McMackin JF (1996) Integrating variable risk preferences, trust, and transaction cost economics. Acad Manage Rev 21(1):73–99

Coulter K, Coulter R (2003) The effects of industry knowledge on the development of trust in service relationships. Int J Res Mark 20(1):31–43

Dyer JH, Chu W (2000) The determinants of trust in supplier-automaker relationships in the U.S., Japan, and Korea. J Int Bus Stud 31(2):259–285

Fawcett SE, Staheli DL, Williams AJ (2006) Creating strategic value through trust: results from practitioner surveys. In: 91st annual international supply management conference

Greenberg PS, Greenberg RH, Antonucci YL (2007) Creating and sustaining trust in virtual teams. Bus Horiz 50(4):325–333

Hartman F (2003) Ten commandments of better contracting: a practical guide to adding value to an enterprise through more effective SMART contracting. ASCE Press, USA

Huang Z, Li SX (2001) Co-op advertising models in manufacturer-retailer supply chains: a game theory approach. Eur J Oper Res 135:527–544

Hung Y-TC, Dennis AR, Robert L (2004) Trust in virtual teams: towards an integrative model of trust formation. In: Proceedings of the 37th Hawaii international conference on system sciences, Hawaii

Kadefors A (2004) Trust in project relationships-inside the black box. Int J Project Manage 22(3):175–182

Kwon I-WG, Suh T (2004) Factors affecting the level of trust and commitment in supply chain relationships. J Supply Chain Manage 40:4–14

Laaksonen T, Jarimo T, Kulmala HI (2009) Cooperative strategies in customer-supplier relationships: the role of interfirm trust. Int J Prod Econ 120(1):79–87

Leverick F, Cooper R (1998) Partnerships in the motor industry: opportunities and risks for suppliers. Long Range Plan 31(1):72–81

Mamdani EH, Assilian S (1975) Experiment in linguistic synthesis with a fuzzy logic controller. Int J Man Mach Stud 7(1):1–13

Mun J, Shin M, Lee K, Jung M (2009) Manufacturing enterprise collaboration based on a goal-oriented fuzzy trust evaluation model in a virtual enterprise. Comput Ind Eng 56:888–901

Panayides PM, Venus Lun Y (2009) The impact of trust on innovativeness and supply chain performance. Int J Prod Econ 122:35–46

Paterson I, Maguire H, Al-Hakim L (2008) Analysing trust as a means of improving the effectiveness of the virtual supply chain. Int J Network Virtual Organ 5(3/4):325–348

Schmidt S, Steele R, Dillon TS, Chang E (2007) Fuzzy trust evaluation and credibility development in multi-agent systems. Appl Soft Comput J 7(2):492–505

Sivanandam SN, Sumathi S, Deepa SN (2007) Introduction to fuzzy logic using Matlab. Springer, Berlin Heidelberg

Tian Y, Lai F, Daniel F (2008) An examination of the nature of trust in logistics outsourcing relationship. Empiric evidence from China. Ind Manage Data Syst 108(3):346–367

Wong PS-P, Cheung S-O (2004) Trust in construction partnering: views from parties of the partnering dance. Int J Project Manage 22:437–446

Zaghloul R, Hartman F (2002) Construction contracts: the cost of mistrust. Int J Project Manage 21(6):419–424

Zaheer A, McEvily B, Perrone V (1998) Does trust matter? Exploring the effects of interorganizational and interpersonal trust on performance. Organ Sci 9(2):141–159

Multi-objective Optimization of a Multi-site Manufacturing Network

Houssem Felfel, Omar Ayadi and Faouzi Masmoudi

Abstract This chapter considers multi-site manufacturing network where multi plants are considered in order to satisfy customer demand. A multi-objective, multi-stage, multi-product, and multi-period model for production and transportation planning in a multi-site manufacturing network is formulated. Two measure criteria, total cost and products' quality level, are simultaneously considered as objective functions to be optimized. The solution of this problem is a set of Pareto fronts that can be used for decision-making. Three optimization method- weighted sum method, epsilon constraint method and goal attainment method- are adapted to solve the considered problem and corresponding results are compared based on an illustrative example. The results show that the epsilon constraint method outperforms the other technique for the considered case.

Keywords Multi-site manufacturing network · Multi-objective optimization · Pareto-optimal solutions

1 Introduction

To cope with the continuous highly competition, manufacturing environments have changed from traditional single-plant to multi-site structure. Multi-site production planning problems have recently attracted many researchers' attention.

H. Felfel (✉) · O. Ayadi · F. Masmoudi
Laboratoire de Mécanique, Mo délisation et Production, Ecole Nationale d'Ingénieurs de Sfax, Université de Sfax, Route de Soukra, BP1173, 3038 Sfax, Tunisie
e-mail: houssem.felfel@gmail.com

O. Ayadi
e-mail: omar.ayadi@yahoo.fr

F. Masmoudi
e-mail: faouzi.masmoudi@enis.rnu.tn

M. S. Abbes et al. (eds.), *Mechatronic Systems: Theory and Applications*, Lecture Notes in Mechanical Engineering, DOI: 10.1007/978-3-319-07170-1_7,

According to (Harland 1997), the multi-site conceptual models can be classified into 4 categories: internal chain (only one plant), dyadic relationship (parallel plants), external chain (serial plants) and network (serial and parallel plants).

In this work, we are particularly interested in multi-site manufacturing network. Vercellis (1999) dealt with a capacitated master production planning and capacity allocation problem for a multi-plant supply network. The author considers a single objective: the minimization of the total cost. To solve this problem, two iterative LP-based heuristic algorithms were developed. Moon et al. (2002) proposed an integrated process planning and scheduling model for the dyadic multi-plant supply chain. The developed model aims to minimize the total using a genetic algorithm-based heuristic approach. Lin and Chen (2007) proposed a monolithic model of the multi-stage multi-site production planning problem of a real thin film transistor-liquid crystal display (TFT-LCD) supply chain network in Taiwan. The developed model combined two different time scales, i.e., monthly and daily time buckets and aims to minimize the total cost. In this chapter we are interested in providing an exact Pareto front to the decision maker. Shah and Ierapetritou (2012) deal with the integrated planning and scheduling multisite, multiproduct problem using the augmented Lagrangian decomposition method. The proposed model minimizes the total costs which include production costs, backorder costs, variable inventory costs, and transportation costs. From the above literature review, multi criteria have been considered insufficiently in multi-site supply network optimization problem.

Many solution methods are proposed in the literature to solve multi-objective optimization problems. These methods can be classified into five major types, including scalar methods, interactive methods, meta heuristic methods, decision aided methods and fuzzy methods (Collette and Siarry 2003). In this chapter, we are interested in providing an exact Pareto front to the decision maker. Among exact methods, we used the weighted sum; the e-constraint method and the goal attainment method in order generate the Pareto optimal solution. To the best of our knowledge, goal attainment method has not been used in the multi-objective supply network optimization.

The remainder of the chapter is organized as follows. The next section is devoted to problem statement. Section 3 details the mathematical formulation of the multi-site manufacturing network planning problem. Section 4 presents the solution approaches. The obtained results are discussed in Sect. 5, followed by the conclusion in Sect. 6.

2 Problem Statement

The multi-site manufacturing network considered in this chapter consists of a manufacturing system wherein the product is processed through different plants. Each production stage may involve more than one plant I_i forming a multi-site manufacturing environment as shown in Fig. 1. A delivery lead time is considered

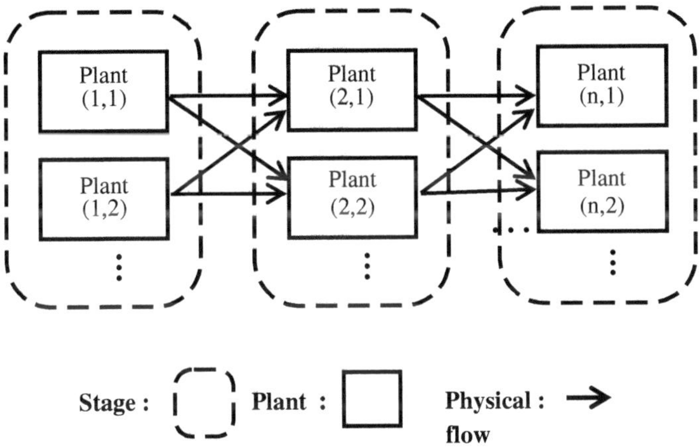

Fig. 1 Multi-site manufacturing network structure

in transportation semi-finished product between upstream and downstream plants. The problem formulated attempts simultaneously to minimize the total system cost and maximizing products quality level. Decision makers in the manufacturing network under concern aim to make the following decisions.

- The production amount in normal and overtime working hours for each production plant in each period from considered time horizon.
- The amount of inventory of each finished or semi-finished product that should be maintained at each plant during the periods.
- The amount of each product to be transported between upstream and downstream plants in each period.

3 Mathematical Formulation

The developed model requires the following indices, parameters and variables.

Indices

L_i Set of direct successor plant of i

ST_j Set of stages (j = 1, 2, …, N)

i, i' Production plant index (i, i' = 1, 2, …, I) where i belongs to stage *n* and i' belongs to stage $n + 1$

k Product index (k = 1, 2, …, K)

t Period index (t = 1, 2, …, T)

Decision variables

P_{ikt}	Production amounts of product k at plant i in period t in normal working hours
H_{ikt}	Production amounts of product k at plant i in period t in overtime
S_{ikt}	Amounts of end of period inventory of product k at plant i in period t
JS_{ikt}	Amounts of end of period inventory of semi-finished product k at plant i in period t
$TR_{i \to i',kt}$	Amounts of product k transported from plant i to i' in period t
$Q_{i',k}$	Amounts of product k received by plant i' in period t

Parameters

cp_{ik}	Unit cost of production for product k in normal working hours at plant i
ch_{ik}	Unit cost of production for product k in overtime at plant i
$ct_{i \to i',k}$	Unit cost of transportation between plant i and i' of production for product k
cs_{ik}	Unit cost of inventory of product or component k at plant i
$capp_{it}$	Production capacity at plant i in normal working hours in period t
$caph_{it}$	Production capacity at plant i in overtime in period t
$caps_{it}$	Storage capacity at plant i in period t
$capt_{i \to i',t}$	Transportation capacity at plant i in period t
D_{kt}	Demand of finished product k in period t
b_k	Time needed for the production of a product entity k [min]
α_{kt}	Quality grade to produce a product k in plant i
DL	Delivery time of the transported quantity
β	Percentage of demand to address the non-quality

Formulation

$$Min\{f(0)\} = \sum_{t=1}^{T}\sum_{k=1}^{K}\sum_{i=1}^{I} cp_{ik}P_{ikt} + ch_{ik}H_{ikt} + cs_{ik*}(S_{ikt} + JS_{ikt}) + ct_{i \to i',k}TR_{i \to i',kt} \tag{1}$$

$$Max\{f(1)\} = \sum_{t=1}^{T}\sum_{k=1}^{K}\sum_{i=1}^{I} \alpha_{ik}(P_{ikt} + H_{ikt}) \tag{2}$$

$$S_{ik,t} = S_{i,kt-1} + P_{ikt} + H_{ikt} - \sum_{i' \in L_i} TR_{i \to i',kt}, \quad \forall i \in ST_{j<N}, \forall k, t \tag{3}$$

$$\sum_{i=1}^{I} S_{ik,t} = \sum_{i=1}^{I} (S_{ik,t-1} + P_{ikt} + H_{ikt}) - D_{kt}, \quad \forall i \in ST_{j=N}, k, t \tag{4}$$

$$JS_{ik,t} = JS_{ik,t-1} + Q_{ikt} - (P_{ikt} + H_{ikt}), \quad \forall i, k, t \tag{5}$$

$$\sum_{t=1}^{T} P_{ikt} + H_{ikt} = \sum_{t=1}^{T} (1 + \beta) D_{kt}, \quad \forall i, k \tag{6}$$

$$Q_{i'k,t+DL} = \sum_{i' \in L_i} TR_{i \to i',kt}, \qquad \forall i', k, t \tag{7}$$

$$\sum_{k=1}^{K} b_k P_{ikt} \leq capp_{it}, \qquad \forall i, t \tag{8}$$

$$\sum_{k=1}^{K} b_k H_{ikt} \leq caph_{it}, \qquad \forall i, t \tag{9}$$

$$\sum_{k=1}^{K} S_{ikt} + JS_{ikt} \leq caps_{it}, \qquad \forall i, t \tag{10}$$

$$\sum_{k=1}^{K} TR_{i \to i',kt} \leq captr_{it} \tag{11}$$

$$P_{ikt}, H_{ikt}, S_{ikt}, JS_{ikt}, TR_{i \to i',kt} \geq 0, \qquad \forall i, k, t \tag{12}$$

The objective function (1) aims to minimize the total costs. The second considered objective function (2) aims to maximize the product of quality level. The Eq. (3) provides the balance for the inventory of products in every production stage except for the last stage. Constraint (4) is the balance equations for the inventory for the last production stage, considering customer demands. Constraint (5) is the inventory balance equation for the semi-finished products. Equation (6) makes sure that the quantity of products should be equal to customer needs taking into account the percentage of demand to address the non-quality. Constraint (7) represents the balance equations for the transportation between the different production stages. Equations (8–11) denotes the capacity constraints. Constraint (12) defines the domain of definition of decision variables.

4 Solution Approaches

4.1 Weighted Sum Method

The weighted sum method is widely used for multi-objective optimization problem. The main idea of this approach is to transform the multi-objective optimization problem into a single objective problem and to associate each objective function with a weighting coefficient (Coello 1999). The weighting factors indicate the relative importance of every corresponding objective functions. The modified problem can be represented as follows:

$$Minimise\, F(x) = \sum_{i=1}^{M} w_i * f_i(x), \quad x \in X.$$

where M the number of objectives w_i *($i = 1.., M$)* is a weighting factor for the ith objective function $f_i(x)$, $w_i \geq 0$ and $\sum_{i=1}^{M} w_i = 1$. By varying the weights of this function, we can obtain different optimal solutions in order to construct the Pareto front.

4.2 The E-constraints Method

The e-constraints method proposed by Chankong and Haimes (1983) is a non-dominated method to generate a set of Pareto solutions for multi-objective problems. The e-constraint method has widely been used in the literature to generate exact Pareto-optimal solutions for multi-objective supply chain planning problem (You and Grossmann 2008; Franca et al. 2010). In the epsilon constraint method, one of the objective functions is selected to be optimized. Then the other objective functions are transformed into constraints by setting an upper bound to each of them. The level of ε_j is then altered in order to generate the entire Pareto optimal set. The modified problem can be represented as follows:

$$\begin{aligned} & Minimise\, f_1(x); \\ & Subject\ to \quad f_j(x) \leq \varepsilon_j \quad \forall j = 2, \ldots M. \\ & x \in X. \end{aligned}$$

4.3 Goal Attainment Method

The goal attainment technique is a variation of goal programming technique used to solve the multi-objective optimization problem (Hwang and Masud 1979). The goal attainment method was applied to solve several real-world multi-objective problems such as production systems (Azaron et al. 2006), project management (Azaron et al. 2007a) and reliability optimization (Azaron et al. 2007b). The preferred solution using this method is fairly sensitive to goal b_i and weight c_i where i = 1.., M. When c_i decreases, the associated objective function becomes more important and should be satisfied. The weight, are generally normalized so that $\sum_{i=1}^{M} c_i = 1$. To generate a set of Pareto-optimal solutions, only one parameter

Fig. 2 Optimal solutions generated by ECM, WSM and GAM

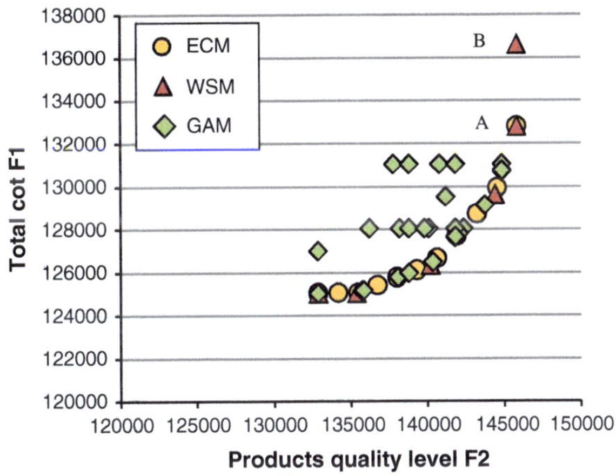

is varied and the others are fixed every time. The modified optimization problem can be represented as follows:

$$\begin{aligned}&\textit{Minimise } z;\\&\textit{Subject to} \quad f_j(x) - c_i * z \leq b_i, \quad \forall i = 1, \ldots, M\\&x \in X.\end{aligned}$$

5 Computational Results

The considered case study consists of 5 production stages with 8 serial and parallel plants and 6 planning period. The weighted sum method, the e-constraint method and goal attainment method were coded using LINGO 14.0 software package. The model example is solved using LINGO 14.0 and MS-Excel 2010 on a 32-bit Windows 7 based computer with an INTEL(R) Core (TM) 2Duo CPU,T5670@1.8 GHZ, 1.8 GHZ, 2 GB RAM. The obtained Pareto optimal solution curve is shown in Fig. 2. Using the weighted sum method, many combinations of weights generate the same points however with the e-constraint method every run can produce a different solution. According to Fig. 2, we notice that the goal attainment method generates many weakly Pareto optimal points i.e. points that are dominated by the other solutions. Furthermore, it can be noted that the weighted sum method gives one weakly efficient solution B dominated by A. However, the e-constraint method generates only efficient Pareto optimal solutions. Concerning computational time, the e-constraint method is more performer than the two other methods since it solves the considered model in 6 s where GAM and WSM necessitate 20 and 33 s respectively and it gives more efficient optimal solutions as shown in Table 1.

Table 1 Computational results of different optimization approaches

Method	CPU time (seconds)	NO. of solutions	NO. of efficient solutions
ECM	6	11	11
GAM	20	31	10
WSM	33	6	5

6 Conclusion

In this chapter, we proposed a multi-objective multi-site manufacturing network programming model by optimizing simultaneously total cost and products quality level. Three multi-objective optimization methods were evaluated in the context of the study: weighted sum method, e-constraints method and goal attainment method. Computational results showed significantly better estimation of the Pareto front by the e-constraint method. In the future works, we will focus on providing the decision maker with a solution from the set of Pareto which satisfies its requirements.

References

Azaron A, Katagiri H, Kato K, Sakawa M (2006) Modelling complex assemblies as a queueing network for lead time control. Eur J Oper Res 174:150–168

Azaron A, Katagiri H, Sakawa M (2007a) Time-cost trade-off via optimal control theory in Markov PERT networks. Ann Oper Res 150:47–64

Azaron Katagiri H, Kato K, Sakawa M (2007b) A multi-objective discrete reliability optimization problem for dissimilar-unit standby systems. OR Spectrum 29:235–257

Chankong V, Haimes Y (1983) Multi-objective decision making theory and methodology. Elsevier Science, New York

Coello CAC (1999) A comprehensive survey of evolutionary-based multi-objective optimization techniques. Knowl Inf Syst Int J 1:269–308

Collette Y, Siarry P (2003) Multi-objective optimization: principles and case studies. Springer, Berlin

Franca RB, Jones EC, Richards CN, Carlson JP (2010) Multi-objective stochastic supply chain modeling to evaluate tradeoffs between profit and quality. Int J Prod Econ 127(2):292–299

Harland C (1997) Supply chain operational performance roles. Integr Manuf Syst 8(2):70–78

Hwang CL, Masud ASM (1979) Multiple objective decision making. Springer, Berlin

Lin JT, Chen YY (2007) A multi-site supply network planning problem considering variable time buckets—a TFT-LCD industry case. Int J Adv Manuf Tech 33(9–10):1031–1044

Moon C, Kim J, Hur S (2002) Integrated process planning and scheduling with minimal total tardiness in multi-plants supply chain. Comput Ind Eng 43:331–349

Shah NK, Ierapetritou MG (2012) Integrated production planning and scheduling optimization of multisite, multiproduct process industry. Comput Chem Eng 37:214

Vercellis C (1999) Multi-plant production planning in capacitated self-configuring two-stage serial systems. Eur J Oper Res 119(2):451–460

You F, Grossmann IE (2008) Design of responsive supply chains under demand uncertainty. Comput Chem Eng 32(12):3090–3111

Influence of Processing Parameters on the Mechanical Behavior of CNTs/Epoxy Nanocomposites

Ayda Bouhamed, Olfa Kanoun and Nghia Trong Dinh

Abstract Due to their extraordinary mechanical and physical properties, carbon nanotubes (CNTs) represent attractive possibilities for developing a potential polymeric composite. This chapter presents in the first part, the fabrication process of epoxy based Multiwalled Carbon Nanotubes (MWCNTs) reinforced composites using solution processing and direct mixing. The variation in the some processing parameters like stirring time, speed of stirring, curing temperature has results in the improvement of mechanical properties like Young's modulus, yield strength, elongation at break. In fact, the change in the preparation methods parameters has a great effect in the mechanical and morphological properties of nanocomposites due to the effective load transfer mechanism and the state of dispersion. The change in properties has been verified by a comparison between the mechanical properties of all samples subjected to a tension. Experimental results reveal that the nanocomposite prepared at (80 °C, 200 rpm, 2 h of stirring) present a higher mechanical properties comparing to the others. In the second part of this chapter, damping behavior of different CNTs/Epoxy nanocomposites samples are investigated by making mechanical hysteresis test in order to constructed the strain—stress curve loaded at different strain to obtain the energy lost.

Keywords Multi-walled carbon nanotubes/epoxy · Mechanical properties · Stirring time · Speed of stirring · Damping · Energy lost · Hysteresis loop

---

A. Bouhamed (✉)
National Engineering School of Sfax, University of Sfax, Sfax, Tunisia
e-mail: ayda.bouhamed@s2013.tu-chemnitz.de

A. Bouhamed · O. Kanoun · N. T. Dinh
Chair for Measurement and Sensor Technology, Chemnitz University of Technology, Chemnitz, Germany

M. S. Abbes et al. (eds.), *Mechatronic Systems: Theory and Applications*,
Lecture Notes in Mechanical Engineering, DOI: 10.1007/978-3-319-07170-1_8,

1 Introduction

Carbon nanotubes (CNTs) are a novel class of one-dimensional nanomaterials with very high expectations and recently ranked as the hottest topic in physics due to the high structural aspect ratio, excellent mechanical properties and superior electrical conductivity (Dalton et al. 2003). Carbon nanotubes are consisting of folded graphene layers with cylindrical hexagonal lattice structure and they are classified in two categories single- walled carbon nanotubes (SWCNTs) contain one tubule and multi-walled carbon nanotubes (MWCNTs) contain coaxial tubules with interlayer 0.34 nm (Dalton et al. 2003; Wagner et al. 1998; Terrones 2003; Ajayan et al. 2000).

The unique mechanical properties of MWCNTs, namely their high strength and stiffness and enormous aspect ratio make them a potential structural element for the improvement of mechanical properties of some structures (Ajayan 1999).

The combination of these properties makes nanotubes appropriate for an extensive range of applications. One of these possible applications is certain their incorporation as reinforcement of polymers matrices in order to develop a novel strong material. But on the other hand, there is a strong interaction among the nanotubes caused by the intermolecular van der Waals interactions and in combination with their high surface area and high aspect ratio, this latter causes significant agglomeration and prevents transfer of their superior properties to the matrix.

Indeed, Huang et al. showed that a good dispersion can be seen at low concentrations of CNT because of diminution of possibility of to re-aggregate another time (Huang and Terentjev 2008). Additionally, high CNTs concentration in polymer matrix increases resin viscosity and makes the dispersion of CNT extremely difficult (Shen et al. 2007).

For those reasons, several approaches have been adopted in oder to obtain intimate mixing of CNTs with the polymer matrix, including direct mixing, solution processing, melt mixing, in situ polymerization and surfactant-assisted mixing. The good dispersion is assured by several mechanical dispersion methods like ultrasonication, calendering, ball milling and shear mixing (Kanagaraj 2010; Choudhary and Gupta 2011).

The main focus in this chapter is to verify the effect of some process parameters on the mechanical behavior of MWCNTs/epoxy nanocomposites fabricated by direct mixing and solution processing method which is a common method for preparing CNT/polymer nanocomposites involves mixing of CNTs and polymer in a suitable solvent. Accordingly, a comparison of mechanical properties will be done between all the nanocomposites samples and a pure epoxy to see the effects and to obtain the optimum conditions of preparation of nanocomposite.

Additionally, an investigation of the hysteresis loop of nanocomposites subjected under loading and unloading cycle for different number of cycle and at different conditions gives an idea about the damping behavior of each nanocomposite.

2 Experiments

Obtaining a uniform CNTs dispersion has been the most significant challenge in this research field. The dispersion is a bigger challenge because of the probability of CNTs clusters formation is high.

In order to get an homogenous dispersion and a good adhesion between CNTs and polymer matrix, a variation of processing parameters like stirring time, speed of stirring, and solvent addition are done to reveal the influence of every parameters. Therefore, we have prepared a pure polymer "epoxy" and six other samples with reinforced polymer by MWCNTs (see Table 1). A fixed percentage content of MWCNTs (0.3 % w) has been used for the preparation of all nanocomposites samples in order to show clearly the effects of varying processing parameters. To prepare the reinforced polymer samples, we have follow this steps as mentioned in the flowchart presented in Fig. 1.

3 Mechanical Characterization

In this part, tensile test is used to reveal the elastic and plastic behavior of each materials and also to measure the mechanical properties. Tensile tests were performed using a Zwick/Roell tensile tester at a crosshead speed of 10 mm/min.

Tensile test is a simple test in which, we subject our samples to uniaxial tension until it fails. The reaction of the different materials is readily recorded and then analyzing data can be done easily.

3.1 Effect of MWCNTs Addition to Epoxy

Our test experiments show that by the addition of CNTs fillers to the epoxy matrix, the tensile strength become two times higher comparing to the neat epoxy which has 39.6 MPa. But the young modulus is decreased. This decrease is explained by the bad dispersion of MWCNT into the epoxy matrix caused by the higher viscosity of CNTs. As consequence, clusters of MWCNTs are still present and cause stress concentration and in consequence decrease on the young modulus of the nanocomposites. The addition of MWCNTs leads to improvement of the epoxy ductility. The elongation percentage increased from 1.2 % for the neat epoxy to 2.3 % for the reinforced epoxy as shown in Fig. 2.

Table 1 Processing parameters of each type of sample

Parameters	S1	S2	S3(untreated CNTs)	S4	S5	S6
Chemical addition	–	–	–	–	–	Acetone
Speed of stirring	200 rpm	200 rpm	200 rpm	1,000 rpm	1,000 rpm	200 rpm
Time of stirring	2 h	2 h	2 h	2 h	5 h	2 h
Temperature	37 °C	80 °C	80 °C	80 °C	80 °C	80 °C

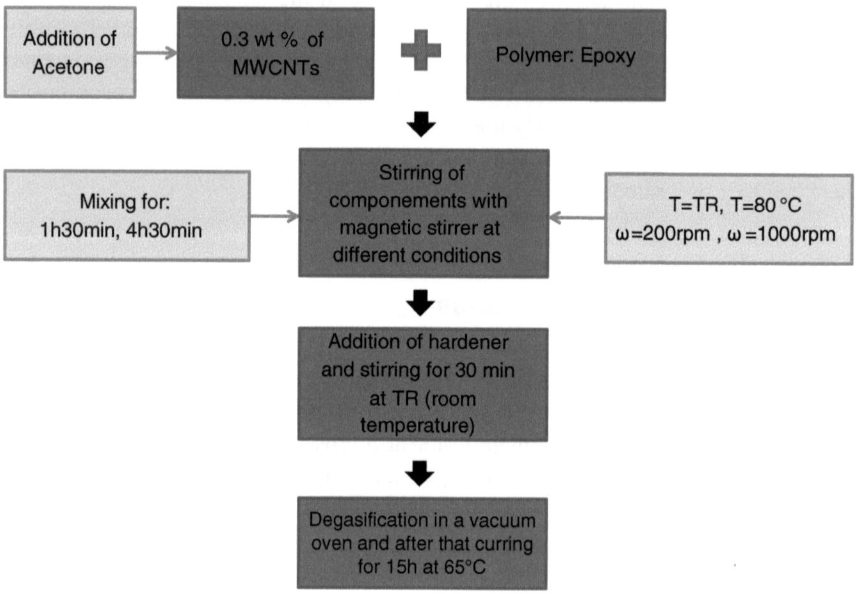

Fig. 1 Flowchart of fabrication process steps of MWCNTs/epoxy nanocomposite including all processing parameters variation

3.2 *Effect of CNTs Surface Modification*

The nanocomposite containing functionalized MWCNTs exhibit a higher mechanical properties than the untreated MWCNT from 3370 to 3460 MPa the elongation at break increased from 3 to 4 % (see Fig. 3). Therefore, surface modification of MWCNTs has enhanced the dispersion contrary to the untreated—MWCNTs presents mainly in the form of agglomerates. The carboxylic functzionalization of MWCNTs improve the interfacial interaction between MWCNTs and epoxy matrix by reducing the hydrophobicity of CNT filler.

Additionally the nanocomposite with functionalized MWCNTs presents better elongation to break as mentioned before this mean that the carboxilic functzionalization of MWCNTs induce better flexibility on the material.

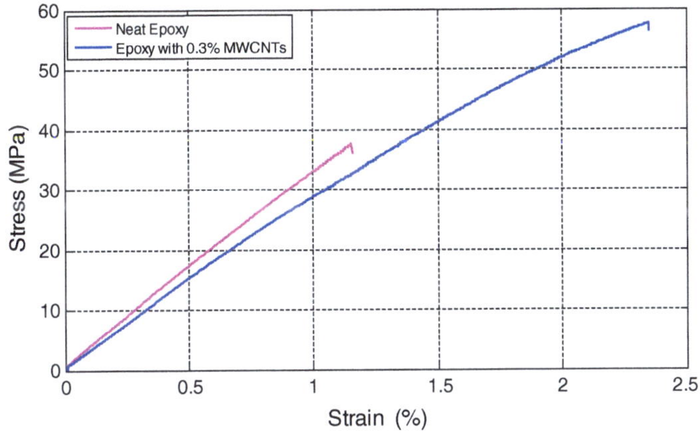

Fig. 2 Mechanical behavior of a neat epoxy and MWCNTs/epoxy polymer nanocomposite prepared at the same conditions

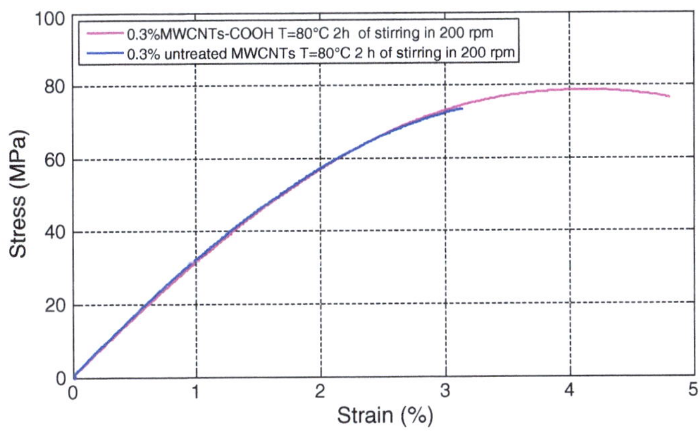

Fig. 3 Mechanical behavior of a Untreated MWCNTs/epoxy polymer nanocomposite and MWCNTs/epoxy polymer nanocomposite prepared at the same conditions

3.3 Effect of Temperature

The nanocomposite prepared at higher temperature presents higher mechanical properties: the tensile strength is two times higher and similarly the elongation at break from 2 to 4 %. At higher temperature, CNTs fillers exhibit increased reactivity with epoxy as a result the dispersion of MWCNTs into the epoxy matrix is facilitates due to the reduction of the viscosity of the polymer. Additionally, the high mixing temperature provides low shear stress which mean low stress concentration as a consequence a diminution of the ability of fibers to fragment and debond (Fig. 4).

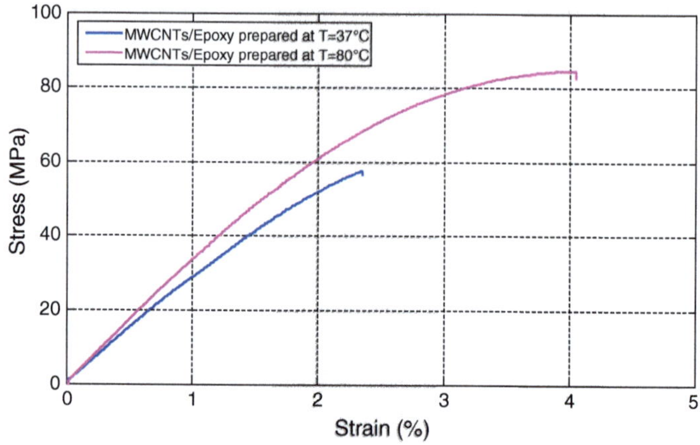

Fig. 4 Mechanical behavior of MWCNTs/epoxy polymer nanocomposite prepared at different temperature

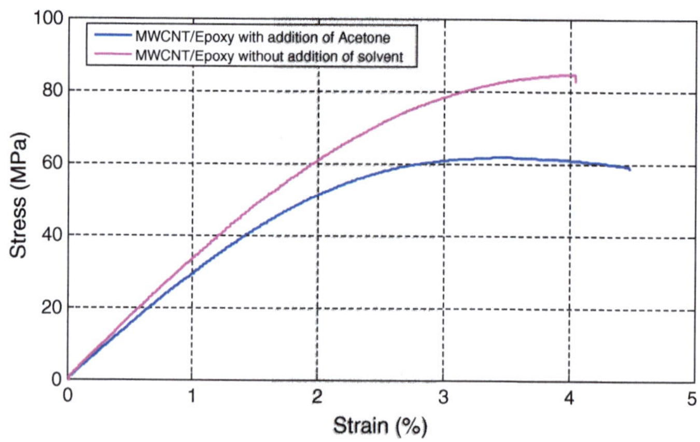

Fig. 5 Mechanical behavior of MWCNTs/epoxy polymer nanocomposite prepared with and without Acetone

3.4 Effect of Solvent Addition

The addition of acetone induces decreasing on the mechanical properties comparing to the nanocomposite prepared at the same conditions and without solvent addition see Fig. 5. The solvent addition increases the dispersion of fillers of CNTs by reducing the resin viscosity, but in contrast, it creates a poor interfacial bonding.

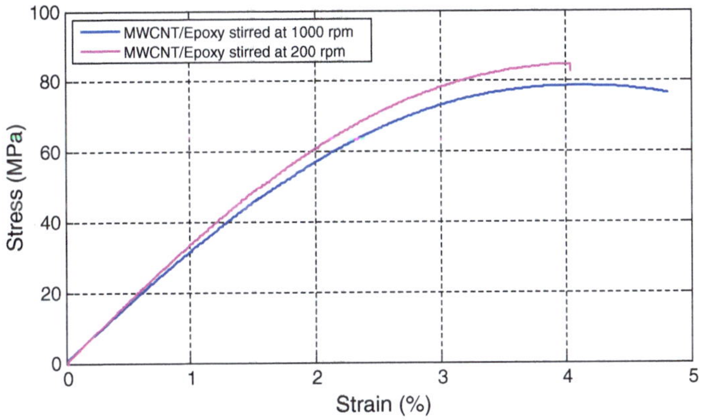

Fig. 6 Mechanical behavior of MWCNTs/epoxy polymer nanocomposite prepared at different stirring speed (1,000 rpm, 200 rpm)

3.5 Effect of Stirring Speed

The nanocomposite stirred at 1,000 rpm exhibit less mechanical properties as it is depicted in Fig.6.

Despite the stirring induce the separation of CNTs clusters by the shear stress which break down the aggregates. Normally this fact lead to uniform dispersion but with higher shear stress, the shear stress imparting on the surface of a CNT can induce a pulling effect (a tensile force) on the nanotube.

As a result, dispersion methods supplying high energy input can also induce fracture of CNTs (Huang and Terentjev 2012).

3.6 Effect of Stirring Time

The nanocomposite prepared for longer time of stirring shows a diminution in the tensile strength and the elongation at break consequence of bad bonding in the interface between the fiber and matrix. More time of stirring after the critical time (the necessary time to separate fibers) create more destroyed CNT fillers, exceptionally; the young modulus is increased because the dispersion of MWCNTs into the epoxy matrix is become more homogenous (Fig. 7).

4 Mechanical Hysteresis

In order to quantify the level of damping in the nanocomposites, dynamic mechanical test is used by subjecting the sample to tension and compression cycles.

The alternation between tension (loading) and compression (unloading) form in hysteresis loop see Fig. 8. The area into the two branches of loading and unloading

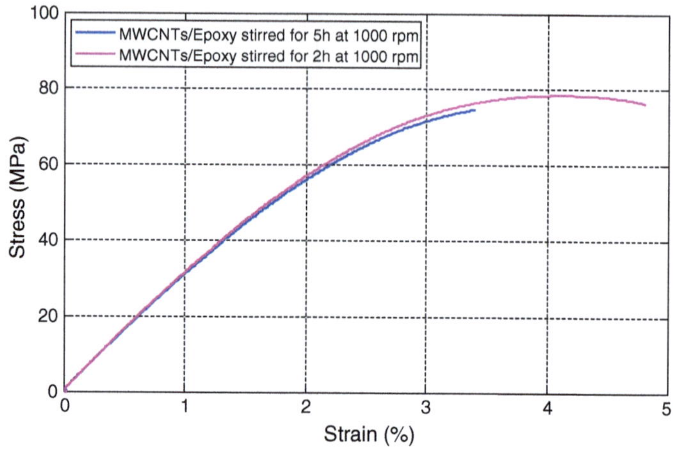

Fig. 7 Mechanical behavior of MWCNTs/epoxy polymer nanocomposite prepared at different stirring time

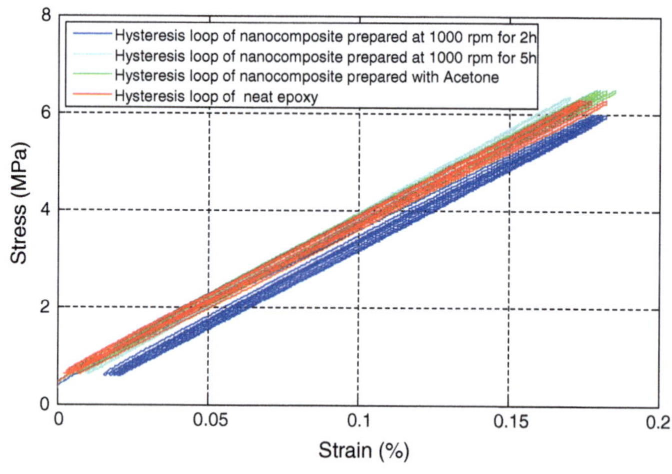

Fig. 8 Hysteresis loops of neat epoxy and different nanocomposites loaded at 0.18 % strain

represent the energy lost. In fact the capacity of damping is proportional to the energy lost or the loop area like is mentioned in Eq. 1 where D is the area enclosed the loop, σ_0, ε_0 are the maximum stress and strain respectively.

$$\eta = \frac{D}{2\sigma_0\varepsilon_0} \tag{1}$$

Our experiment test is done by stretching and restretching the sample under a stable load for in 5 cycles and after that we increase load.

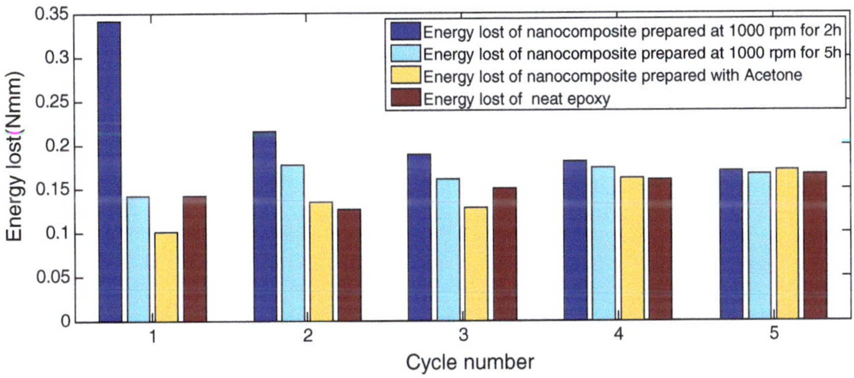

Fig. 9 Energy lost per cycle for a sample loaded at 0.18 % strain

In the Fig. 9, a presentation of the energy lost per cycle for each cycle. In the first test all samples are loaded at 0.18 % strain, the energy lost of the materials is increased comparing with the neat matrix (≈0.1287 Nmm). In fact the addition of the fillers of MWCNTs increase the energy lost for example prepared at 1,000 rpm for 2 h of stirring (≈0.1894 Nmm), in consequence, the damping of nanocomposites samples are also raised, which is related to the poor fiber-matrix adhesion and as results the interface is start to be destroyed in a short time.

Inconsequence more energy is converted to heat and thus a region of potentially high damping.

At higher levels of cyclic strain as shown in Fig. 10, a great amount of heat is generated. As result, the energy lost is become about 4 times bigger comparing to the samples loaded at 0.18 % strain, this is a result of much degradation in the fiber matrix interfaces for all samples especially for nanocomposite prepared with acetone which dissipate energy more than the neat epoxy. Indeed, this result is certainly related to the imperfect interfacial bonding created by the addition of solvent. In fact, the poor adhesion of fibers to matrix has presented an intermediate for the mobility of polymer chains in composites, this explain the higher damping. While the others samples dissipate less energy comparing to the neat epoxy because of the reduction of the mobility of epoxy matrix around CNTs due to the enhancement of the interfacial bonding created by carboxylic functionalization of CNTs.

The exception is observed for the nanocomposite prepared at 1,000 rpm for 5 h which has approximately the same energy lost between (0.15 and 0.2 Nmm) like in the case of sample loaded at 0.18 % strain and as consequence the shape of the hysteresis loop become linear. Hence the material tends to be elastic and also the material has a capability to balance the generated heat by the surrounding environment.

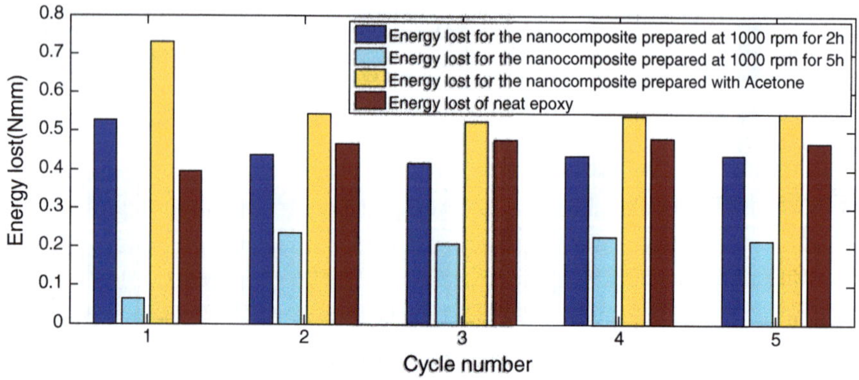

Fig. 10 Energy lost per cycle for a sample loaded at 0.38 % strain

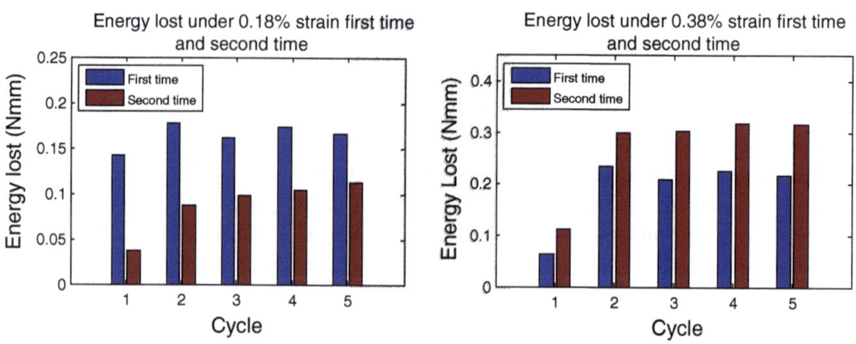

Fig. 11 Effect of repetition of the same test for second time on the energy lost for different loads

Under those circumstances, the damping of the material prepared at 1,000 rpm for 5 h is reduced because of the highest degree of homogeneity in the fiber-matrix distribution induced by the longer time of stirring, that create an increasing in thermal stability at elevated temperature.

Moreover, restretching the sample prepared at 1,000 rpm for 5 h for a second time under the same loads shows that for a small load the energy lost is decreased comparing to the first time, but by increasing the load, the material start to degraded; the interface delamination between fiber and matrix is rising progressively and then more heat is lost and a higher damping comparing to the first time (see Fig. 11).

5 Conclusion

To conclude, the addition of nanotubes has enhanced the tensile properties of all nanocomposites comparing to the neat epoxy, but the degree of improvement is related to processing parameters.

First of all, modifying the surface of CNTs by integrating functional groups is a primary condition before using it in any application.

By increasing the temperature, the materials show better mechanical properties. In fact, increasing temperature plays an important role in the homogeneity of the structure and to get a better interfacial bonding between CNTs and the matrix by the fact of increasing the mobility of polymer to recover the surfaces of the CNTs.

Additionally, stirring has a great effect on the distribution of CNTs but it must be carried out carefully by taken into consideration the contents of CNTs in order to prevent their destruction and reduce their reinforcing effects by making the mixture in optimal speed and optimal time of stirring.

Eventually, the damping is related to both crucial factors which are the interaction between the matrix and the fibers; as a consequence, the hysteresis loss energy increases as interface shear stress decreases when interface is debonded and the second factor is the degree of fiber matrix homogeneity.

Acknowledgements This work is partially supported by the chair for Measurement and Sensor Technology in Chemnitz University of Technology, Germany. The authors also gratefully acknowledge the helpful comments and suggestions of the reviewers, which have improved the presentation.

References

Ajayan PM (1999) Nanotubes from Carbon. Chem Rev 99(7):1787–1799

Ajayan PM, Schadler LS, Giannaris C, Rubio A (2000) Single-walled carbon nanotube–polymer composites: strength and weakness. Adv Mater 12:750–753

Choudhary V, Gupta A (2011) Polymer/Carbon Nanotube Nanocomposites. Chap. 4, Carbon 215 nanotubes-Polymer Nanocomposites, doi:10.5772/18423

Dalton AB, Collins S, Muñoz E, Razal JM, Ebron VH, Ferraris JP, Coleman JN, Kim BG, Baughman RH (2003) Super-tough carbon nanotube fibres. Nature 423:703–703. doi:10.1038/423703a

Huang YY, Terentjev EM (2008) Dispersion and rheology of carbon nanotubes in polymers. Int J Mater Form 63–74

Huang YY, Terentjev EM (2012) Dispersion of Carbon Nanotubes: mixing, sonication, stabilization, and composite properties. Polymers 4:275–295. doi:10.3390/polym4010275

Kanagaraj S (2010) Polyetyethylene Nanotube Nanocomposite, Chap. 5, Polymer Nanotube Nanocomposites: synthesis, properties, and applications, pp 113–136

Shen J, Huang W, Wu L, Hu Y, Ye M (2007) The reinforcement role of different amino-functionalized multi-walled carbon nanotubes in epoxy nanocomposites. Compos. Sci. Technol, 3041–3050

Terrones M (2003) Science and technology of the twenty-first century: synthesis, properties, and applications of carbon nanotubes. Ann Rev Mater Res 33:419–501. doi: 10.1146/annurev.matsci.33.012802.100255

Wagner HD, Lourie O, Feldman Y, Tenne R (1998) Stress-induced fragmentation of multiwall carbon nanotubes in a polymer matrix. Appl Phys Lett 72:188–190

A Polynomial Chaos Method for the Analysis of Uncertain Spur Gear System

Ahmed Guerine, Yassine Driss, Moez Beyaoui, Lassaad Walha, Tahar Fakhfakh and Mohamed Haddar

Abstract This chapter presents a dynamic behavior of a spur gear system with uncertainty associated to friction coefficient. Moreover, the friction coefficient admits some dispersion due to the manufacturing processes. Therefore, it becomes necessary to take this uncertainty into account in the stability analysis of a spur gear system. The proposed method proves to be interesting alternative to the classic methods such as parametric studies. Polynomial chaos approach is a more efficient probabilistic tool for uncertainty propagation.

Keywords Friction coefficient · Uncertainty · Chaos polynomial

A. Guerine (✉) · Y. Driss · M. Beyaoui · L. Walha · T. Fakhfakh · M. Haddar
Mechanics, Modelling and Manufacturing Laboratory LA2MP, Mechanical Engineering Department, National School of Engineers of Sfax, BP 1173, 3038 Sfax, Tunisia
e-mail: ahmedguerine@gmail.com

Y. Driss
e-mail: drissyassine@yahoo.fr

M. Beyaoui
e-mail: moez.beyaoui@yahoo.fr

L. Walha
e-mail: walhalassaad@yahoo.fr

T. Fakhfakh
e-mail: Tahar.Fakhfakh@enis.rnu.tn

M. Haddar
e-mail: mohamed.haddar@enis.rnu.tn

M. S. Abbes et al. (eds.), *Mechatronic Systems: Theory and Applications*, Lecture Notes in Mechanical Engineering, DOI: 10.1007/978-3-319-07170-1_9,

1 Introduction

The spur gear system is characterized by the presence of friction coefficient that affects the vibration of the mechanisms of transmission by gear tooth. These systems play an important role in industrial applications. Some parametric studies have presented the sensitivity of the dynamic behavior of a spur gear system in the presence of a friction coefficient. However, the friction admits some dispersion (Nechak et al. 2011). Therefore, it becomes necessary to take into account these uncertainties in order to study the robustness of a gear system.

Different approaches have been proposed in the literature. The polynomial chaos method can be seen as an interesting alternative. The efficiency of its intrusive and non-intrusive approach has been presented in more applications such as treating uncertainties in multibody dynamic systems (Sandu et al. 2006), environmental and biological problems (Isukapalli et al. 1998) and robust analysis of uncertain non-linear dynamic systems (Nechak et al. 2010). Fischer and Bhattacharya propose the generalized polynomial chaos method to study the stability and instability of stochastic nonlinear dynamic systems (Fisher and Bhattacharya 2008). The main idea of their method is to transform the stochastic differential equations by means of an intrusive Galerkin projection into a deterministic set of differential equations.

The main originality in this chapter is that the friction coefficient between teeth in contact is taken into account in the dynamic behavior study of a gear system. The one stage gear system was modeled by eight-degrees of freedom. We apply a polynomial chaos approach to obtain a system of dynamic equations corresponding to an uncertain friction coefficient. Then we intend to analyze the robustness of a gear system by studying its dynamic response.

2 Plane Description of a One-Stage Gear System

The dynamic model of the one stage gear system is represented on Fig. 1. This model is composed of two blocks. Every block is supported by flexible bearing which the bending stiffness is k_1^x and the traction-compression stiffness is k_1^y for the first block, k_2^x and k_2^y for the second block. The two shafts (1) and (2) admit some torsional stiffness k_1^θ and k_2^θ.

Wheels (11) and (22) characterize respectively the motor side and the receiving machine side. Angular displacements of every wheel are noticed by $\theta(1,1)$, $\theta(1,2)$, $\theta(2,1)$ and $\theta(2,2)$. Besides, the linear displacements of the bearing noted by x_1 and y_1 for the first block, x_2 and y_2 for the second block, are measured in the plan which is orthogonal to the wheels axis of rotation (Kahraman et al. 2007).

Generally we can modelled the gearmesh stiffness variation k(t) by a square wave which is the most representative of the real phenomenon (Fig. 2).

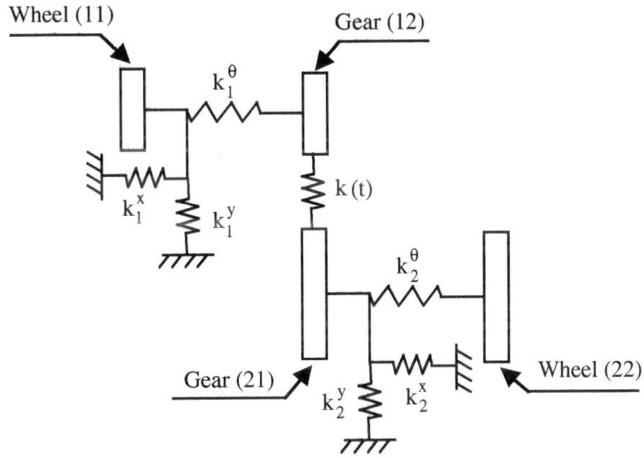

Fig. 1 Global dynamic model of the one stage gear system

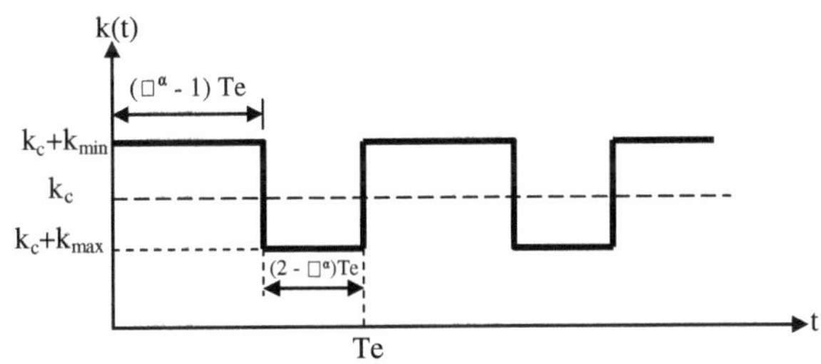

Fig. 2 Modelling of the mesh stiffnesses variation

The gearmesh stiffness variation can be decomposed in two components: an average component noted by k_c, and a time variant one noted by kv(t) (Walha et al. 2009).

3 Friction Coefficient Modelling

During time, there are alternative take between one to two pairs of teeth in contact which causes the birth of single or two friction forces. The modeling of friction forces will be made a comprehensive manner passing by Coulomb's law simplified with a constant coefficient of friction. In the dynamic model, the friction can be introduced by two friction torques applied on the gears (12) and (21).

The friction torque applied on the gear (12) is expressed by:

$$Cf_{12}(t) = F_f(t) \cdot \xi_1(t) \tag{1}$$

Also, the friction torque applied on the gear (21) is expressed by:

$$Cf_{21}(t) = F_f(t) \cdot \xi_1(t) \tag{2}$$

$\xi 1(t)$ and $\xi 2(t)$ are the time varying length between the pitch point P and the corresponding center point of the gears (12) and (21) respectively.

The effort of friction at first pair in contact is defined by:

$$F_f^1(t) = \mu \cdot k^1(t) \cdot \delta(t) \quad (3)$$

The effort of friction at the second pair in contact is defined by:

$$F_f^0(t) = \mu \cdot k^0(t) \cdot \delta(t) \quad (4)$$

$\delta(t)$ is the deflection along the gearmesh contact, it can be written by:

$$\delta(t) = \sin(\alpha) \cdot (x_1 - x_2) + \cos(\alpha) \cdot (y_1 - y_2) + r^b_{(1,2)}\theta_{(1,2)} - r^b_{(2,1)}\theta_{(2,1)} \quad (5)$$

4 Equations of Motion

The set of equations describing the motion of the dynamic model with eight-degrees of freedom of a simple transmission represented in Fig. 1 is defined by:

$$m_1\ddot{x}_1 + k_1^x x_1 + \sin(\alpha)\, k(t) \left\langle L^\delta \right\rangle \{Q\} = \left(F_f^1 - F_f^0\right) \sin(\alpha) \quad (6)$$

$$m_1\ddot{y}_1 + k_1^y y_1 + \cos(\alpha)\, k(t) \left\langle L^\delta \right\rangle \{Q\} = -\left(F_f^1 - F_f^0\right) \cos(\alpha) \quad (7)$$

$$m_2\ddot{x}_2 + k_2^x x_2 - \sin(\alpha)\, k(t) \left\langle L^\delta \right\rangle \{Q\} = -\left(F_f^1 - F_f^0\right) \sin(\alpha) \quad (8)$$

$$m_2\ddot{y}_2 + k_2^y y_2 - \cos(\alpha)\, k(t) \left\langle L^\delta \right\rangle \{Q\} = \left(F_f^1 - F_f^0\right) \cos(\alpha) \quad (9)$$

$$I_{(1,1)}\ddot{\theta}_{(1,1)} + k_1^\theta \left(\theta_{(1,1)} - \theta_{(1,2)}\right) = C_m \quad (10)$$

$$I_{(1,2)}\ddot{\theta}_{(1,2)} - k_1^\theta \left(\theta_{(1,1)} - \theta_{(1,2)}\right) + r^b_{(1,2)}\, k(t) \left\langle L^\delta \right\rangle \{Q\} = Cf^1_{12}(t) - Cf^0_{12}(t) \quad (11)$$

$$I_{(2,1)}\ddot{\theta}_{(2,1)} - k_2^\theta \left(\theta_{(2,1)} - \theta_{(2,2)}\right) - r^b_{(2,1)}\, k(t) \left\langle L^\delta \right\rangle \{Q\} = -Cf^1_{21}(t) + Cf^0_{21}(t) \quad (12)$$

$$I_{(2,2)}\ddot{\theta}_{(2,2)} + k_2^\theta \left(\theta_{(2,1)} - \theta_{(2,2)}\right) = 0 \quad (13)$$

Where the torque friction components are expressed by:

$$Cf^p_{12}(t) = \mu \cdot k^p(t) \cdot \delta(t) \cdot \xi_1^p(t) \text{ and } Cf^p_{21}(t) = \mu \cdot k^p(t) \cdot \delta(t) \cdot \xi_2^p(t) \quad (14)$$

p represents the case of the pair 0 or 1.

$\langle L^{\delta} \rangle$ is defined by:

$$\langle L^{\delta} \rangle = [\sin(\alpha) \quad -\sin(\alpha) \quad \cos(\alpha) \quad -\cos(\alpha) \quad 0 \quad r^{b}_{(1,2)} \quad -r^{b}_{(2,1)} \quad 0] \tag{15}$$

α is the pressure angle equal to 20° and $r^{b}_{(1,2)}$, $r^{b}_{(2,1)}$ represent the base gears radius.

{Q} is the vector of the model generalized coordinates, it is in the form:

$$\{Q\} = [x_1 \quad y_1 \quad x_2 \quad y_2 \quad \theta_{(1,1)} \quad \theta_{(1,2)} \quad \theta_{(2,1)} \quad \theta_{(2,2)}]^{T} \tag{16}$$

5 Application of Polynomial Chaos

The friction coefficient μ is assumed governed by a uniform distribution law on a dispersion interval [a b]. It can be expressed by a uniform stochastic variable $\xi \in [-1\,1]$:

$$\mu = \frac{a+b}{2} + \frac{b-a}{2}\xi \tag{17}$$

The results corresponding to the interval $[a\ b] = [00.1]$ are shown below.

Considering $x_1 = z_1$, $\dot{x}_1 = z_2$, $y_1 = z_3$, $\dot{y}_1 = z_4$, $x_2 = z_5$, $\dot{x}_2 = z_6$, $y_2 = z_7$, $\dot{y}_2 = z_8$, $\theta_{(1,1)} = z_9$, $\dot{\theta}_{(1,1)} = z_{10}$, $\theta_{(1,2)} = z_{11}$, $\dot{\theta}_{(1,2)} = z_{12}$, $\theta_{(2,1)} = z_{13}$, $\dot{\theta}_{(2,1)} = z_{14}$, $\theta_{(2,2)} = z_{15}$ and $\dot{\theta}_{(2,2)} = z_{16}$ (18)

The equations of motion are expressed with a state space representation as:

$$\dot{z}(t) = A(\mu)z(t) + f_{NL}(z(t), \mu) \tag{19}$$

Where

$$\begin{aligned} z(t) = [&z_1(t)\ z_2(t)\ z_3(t)\ z_4(t)\ z_5(t)\ z_6(t)\ z_7(t)\ z_8(t)\ z_9(t) \\ &z_{10}(t)\ z_{11}(t)\ z_{12}(t)\ z_{13}(t)\ z_{14}(t)\ z_{15}(t)\ z_{16}(t)]^{T} \end{aligned} \tag{20}$$

The state variables of the model studied are random processes that may be expressed in the base of the Legendre polynomial chaos: (Jakerman and Roberts 2009)

$$z_i(t, \xi) = \sum_{j=0}^{p} \bar{z}_{i,j}(t) L_j(\xi) \tag{21}$$

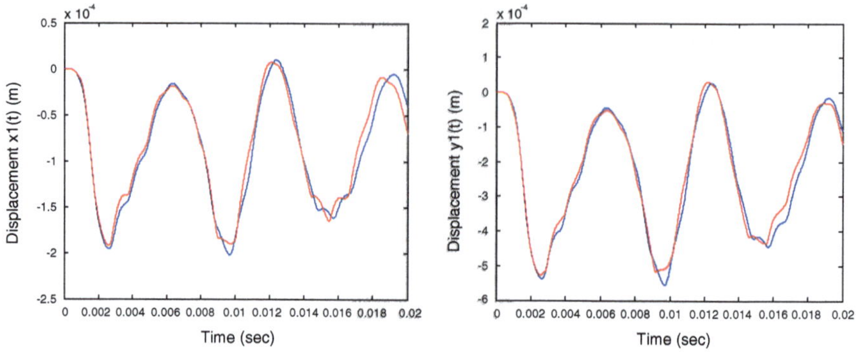

Fig. 3 Temporal fluctuation of the displacement of the first bearing (—: μ = 0; – –: μ = 0.05)

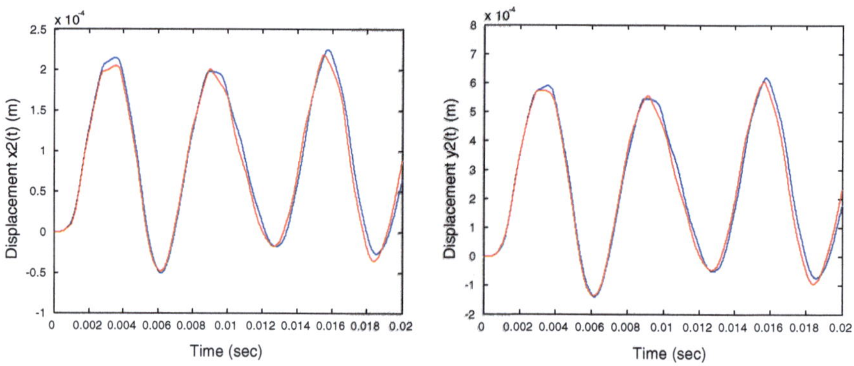

Fig. 4 Temporal fluctuation of the displacement of the second bearing (—: μ = 0; – –: μ = 0.05)

6 Numerical Analysis

We represent in the Fig. 3 the temporal response of the dynamic component of the displacement at the first bearing in the directions x and y. We show that the presence of friction affects little the linear displacements of the bearing without changing the general form of the signal. This result is the same on the second bearing. We represent on Fig. 4 the temporal response of the dynamic component of the displacement at the second bearing in two directions x and y.

The Figs. 5 and 6 show the effects of friction on the dynamic behavior of the wheels. By analyzing the angular fluctuations of driving and driven wheel, the effects of friction are clearly identified. From these figures we notice that the presence of friction between teeth affect the angular velocity of wheel that approaching a zero velocity corresponding to the stop of wheel rotation. The

Fig. 5 Temporal fluctuations of the angular velocity of the wheel θ(1,1) (—: μ = 0; – –: μ = 0.05)

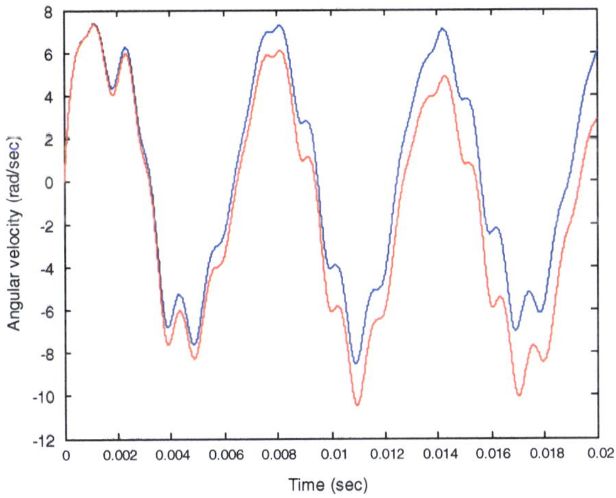

Fig. 6 Temporal fluctuations of the angular velocity of the wheel θ(2,2) (—: μ = 0; – –: μ = 0.05)

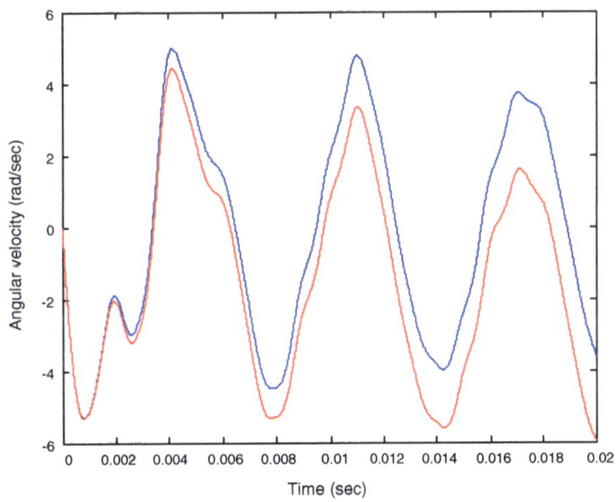

increase of the coefficient of friction reduced more the velocity and accelerates the stop of wheel rotation.

If we take about uncertainty, we can represent, on Fig. 7, the instantaneous mean value and variance of the linear displacement of the first bearing following the direction x. we notice that the amplitude of this displacement and variance is very low. The amplitude is of the order of 10^{-4} m. Also we can find that the mean value and variance of this displacement is periodic and fluctuate around a zeros value. We represent on Fig. 8 the instantaneous mean value and variance of the linear displacement of the second bearing following the direction y.

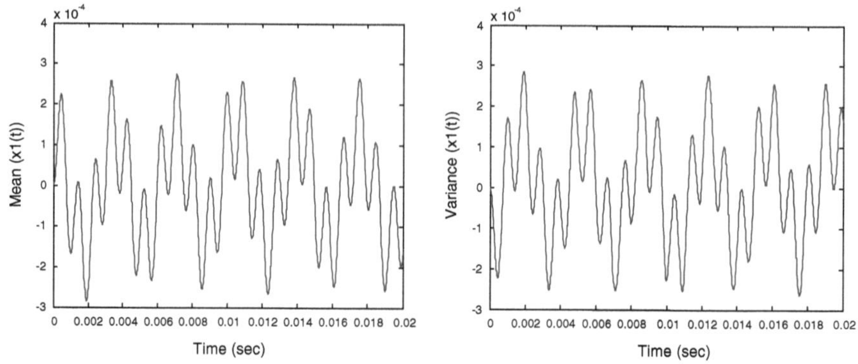

Fig. 7 Instantaneous mean value and variance of $x_1(t)$

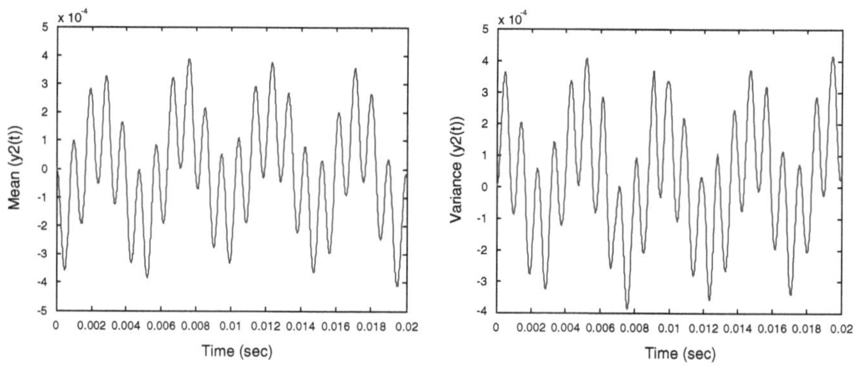

Fig. 8 Instantaneous mean value and variance of $y_2(t)$

7 Conclusion

This chapter has presented the problem of the global dynamic model of the one stage gear system which proves important when parametric uncertainty is friction coefficient. An approach based on the polynomial chaos theory has been proposed to study the dynamic behavior of this system which in the presence of friction which is highlighted by an uncertain coefficient. A complete study of the dynamic behavior including stability and vibratory analyses has been carried out for eight-degrees of freedom model characterized by an uncertain friction coefficient.

The main results of the present study show that the polynomial chaos may be an efficient tool to take into account the dispersions of the friction coefficient in the dynamic behavior study of one-stage gear system.

References

Fisher J, Bhattacharya R (2008) Stability analysis of stochastic systems using polynomial chaos. In: American control conference, USA, pp 4250–4255, 11–13 June

Isukapalli SS, Roy A, Georgopoulos PG (1998) Stochastic response surface methods (SRSMs) for uncertainty propagation: application to environmental and biological systems. Risk Anal 18:351–363

Jakerman JD, Roberts SG (2009) Stochastic galerkin and collocation methods for quantifying uncertainties in differential equation: a review. ANZIAM J 50:815–830

Kahraman A, Lim J, Ding H (2007) A dynamic model of a Spur gear pair with Friction. In: 12th IFToMM world congress, Besançon (France) 18–21 June 2007

Nechak L, Berger S, Aubry E (2010) Robust analysis of uncertain dynamic systems: combination of the centre manifold and polynomial chaos theories. WSEAS Trans Syst 9:386–395

Nechak L, Berger S, Aubry E (2011) A polynomial chaos approach to the robust analysis of the dynamic behaviour of friction systems. Eur J Mech A Solids 30:594–607

Sandu A, Sandu C, Ahmadian M (2006) Modeling multibody dynamic systems with uncertainties—part I: numerical application. Multibody Syst Dyn 15:369–391

Walha L, Fakhfakh T, Haddar M (2009) Nonlinear dynamics of a two-stage gear system with mesh stiffness fluctuation, bearing flexibility and backlash. Mech Mach Theory 44:1058–1069

Non-linear Stiffness and Damping Coefficients Effect on a High Speed AMB Spindle in Peripheral Milling

Amel Bouaziz, Maher Barkallah, Slim Bouaziz, Jean-Yves Cholley and Mohamed Haddar

Abstract During the milling operation, some factors such as the cutting parameters and the spindle type and speed can influence the cutting force attitude consequently the state of the finished part. In this work, a High Speed Milling (HSM) spindle supported by a pair of Active Magnetic Bearings (AMB) is modeled. The shaft is discretized with Timoshenko beam finites elements. The six degrees of freedom of both rigid and elastic motions are considered. Electromagnetic forces are modeled by linear springs and dampers. A peripheral milling model is suggested to predict the dynamic cutting force and the tool tip response. Dynamic coefficients of bearings with four, six and eight electromagnets are considered and plotted to study the spindle dynamic behavior.

Keywords Milling · Cutting parameters · Non-linear · Dynamic coefficients

A. Bouaziz (✉) · M. Barkallah · S. Bouaziz · M. Haddar
Mechanics Modeling and Production Laboratory (LA2MP), National School of Engineers of Sfax (ENIS), University of Sfax Tunisia, BP 1173, 3038 Sfax, Tunisia
e-mail: amelkharrat@yahoo.fr

M. Barkallah
e-mail: bark_maher@yahoo.fr

S. Bouaziz
e-mail: slim.bouaziz1@gmail.com

M. Haddar
e-mail: Mohamed.haddar@enis.rnu.tn

J.-Y. Cholley
Laboratory of Engineering in Mechanical Systems and Materials (LISMMA), SUPMECA, Saint-Ouen, France
e-mail: jean-yves.choley@supmeca.fr

M. S. Abbes et al. (eds.), *Mechatronic Systems: Theory and Applications*,
Lecture Notes in Mechanical Engineering, DOI: 10.1007/978-3-319-07170-1_10,

1 Introduction

High speed milling at speeds beyond 10,000 rpm is rapidly growing in many manufacturing industries such as automotive, aerospace, moulds, etc. So, reaching the best cutting at such high speeds requires the well studying of the cutting parameters as well as the spindle system. Many researches were produced to study milling spindle especially AMB. In fact Knospe (2007) showed that AMB are characterized by their accuracy, high robustness to shock and their high rotational speed proved also in studies done by Kimman et al. (2010) and Gourc et al. (2011). Concerning the AMB components (Bouaziz et al. 2011, 2013) studied the impact of the electromagnet number variation on a rigid rotor dynamic behaviour and they showed that the increase of the electromagnet number lets reducing the vibratory level. In addition Belhadj Messaoud et al. (2011) presented the effect of the air gap and the rotor speed on the electromagnetic forces. They concluded that the intensity of electromagnetic forces increases with the decrease of the air gap. Also, among the AMB benefits, their components such as sensors and feedback currents can be used to predict the cutting forces. In fact Auchet et al. (2004) presented a new method for measuring the cutting forces by analyzing the command voltages of AMB. Lai (2000) studied the influence of dynamic radii, cutting feed rate, and radial and axial depths of cut on milling forces. He found that chip thickness has the most significant influence on the forces. In the same context Klocke et al. (2009) investigated the influence of speed and feed per tooth in micro milling on the surface quality and tool life. They showed that feed rate have an important influence on the surface quality. In fact, to increase the surface quality, it is necessary to decrease the feed rate value. Budak (2006a, b) presented the milling force, the part and tool deflection, form error and the stability models. From this method, he can check the process constraints and select optimal cutting conditions. Concerning the machining process stability, Faassen et al. (2003) expanded a dynamic model for the milling process in which the stability lobes have been generated. This model predicts the stability limit slightly too conservative. In this chapter, a HSM spindle with AMB is modeled by the finite element method based on the Timoshenko beam theory. Rigid displacements are also taken into account (Hentati et al. 2013). Electromagnetic forces are modeled by spring and damping coefficients. Peripheral milling process is modeled and cutting forces are formulated. The dynamic response at tool-tip, the cutting force attitude is plotted. The dynamic coefficients of the used AMB are then investigated.

2 Spindle Modeling

In the studied spindle modeling, the shaft is discretised into 23 finites elements with two nodes and six degrees of freedom. The modeling is also based on counting the rigid displacements. The classical bearing configuration is made in the modeling. Two AMB and an axial bearing are used to suspend the rotor in the central position. The resultant radial force of such pair of electromagnets is a nonlinear function of the currents, rotor position, and magnetization of the iron core. It is expressed according to Bouaziz et al. (2011) as follow:

$$f_j(I_j, u_j) = -a\left(\left[\frac{I_0 - I_j}{e_0 - u_j}\right]^2 - \left[\frac{I_0 + I_j}{e_0 + u_j}\right]^2\right) \quad j = x, y \tag{1}$$

u_j presents the small deformations in the j direction and e_0 is the nominal air gap between the shaft and the stator.

a is the global magnetic permeability and it is expressed as follow:

$$a = \frac{\mu_0 S n^2 I_0^2}{4} \cos\theta \tag{2}$$

μ_0, n, S, I_0 and θ are respectively the vacuum permeability, the windings number, the cross sectional area, the bias current and the half angle between the poles of electromagnets. A proportional-derivative (PD) controller is used to determine the control current I_j is written as $I_j = K_p u_j + K_p \dot{u}_j$ (Bouaziz et al. 2011).

$\dot{u}_j$ are velocities in the j direction. The three types of AMB, where four, six and eight electromagnets are presented in (Fig. 1).

As shown in Fig. 2, the electromagnetic field is modeled by stiffness and damping coefficients.

The resulting electromagnetic forces at each bearing as non-linear function of the control current and the rotor displacements and velocities $(u_j, \dot{u}_j)$ can be written in matrix form as follow (Belhadj Messaoud et al. 2011):

$$[K_{ij}]\begin{Bmatrix} u_x \\ u_y \end{Bmatrix} + [C_{ij}]\begin{Bmatrix} \dot{u}_x \\ \dot{u}_y \end{Bmatrix} = \begin{Bmatrix} f_x \\ f_y \end{Bmatrix} \tag{3}$$

$$[K_{ij}] \text{ is the stiffness matrix : } [K_{ij}] = \begin{bmatrix} K_{xx} & K_{xy} \\ K_{yx} & K_{yy} \end{bmatrix} \tag{4}$$

$$[C_{ij}] \text{ is the damping matrix : } [C_{ij}] = \begin{bmatrix} C_{xx} & C_{xy} \\ C_{yx} & C_{yy} \end{bmatrix} \tag{5}$$

According to Kimman et al. (2010), the axial bearing force is linearized as:

$$f_z = K_{iz} I_z + K_z u_z \tag{6}$$

K_{iz}, I_z, K_z and u_z are respectively, the force current dependency, the control current of the actuator, the negative stiffness of the axial bearing and the vertical displacement of the rotor.

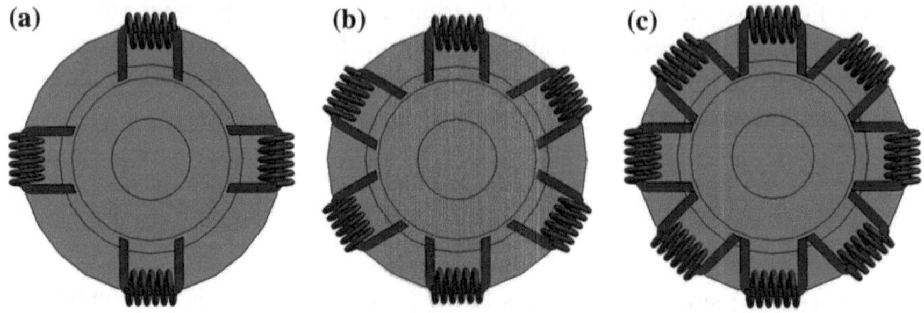

Fig. 1 AMB models. **a** 4 electromagnets, **b** 6 electromagnets, **c** 8 electromagnets

Fig. 2 AMB modeling by non-linear spring and damping coefficients

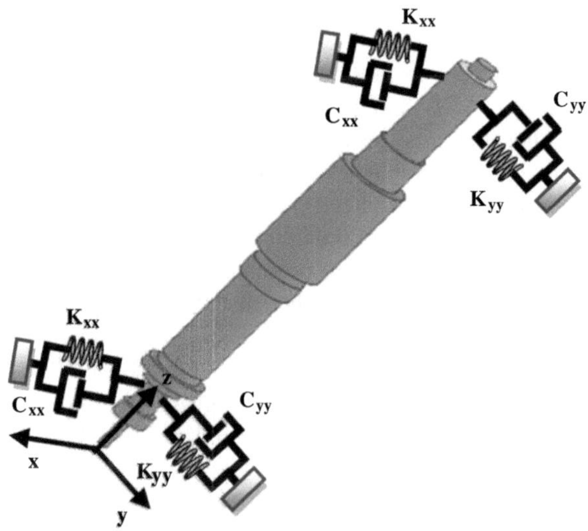

3 Cutting Force Model

A dynamic cutting force model of peripheral milling is developed (Fig. 3). According to this configuration, cutting force is composed of a tangential component F_t orthogonal to the specific segment of the cutting edge. This force is proportional to the chip thickness and the axial depth of cut. A radial component F_r is proportional to F_t and orthogonal to both the cutting edge segment and the z- axis. The third component is the axial force F_a. F_t F_r and F_a can be calculated as function of the instantaneous chip thickness $H(\phi_j(t))$:

$$\begin{cases} F_t = K_t a_p H(\phi_j(t)) \\ F_r = K_r F_t \\ F_a = K_a F \end{cases} \tag{7}$$

Fig. 3 Modeling of the peripheral milling process

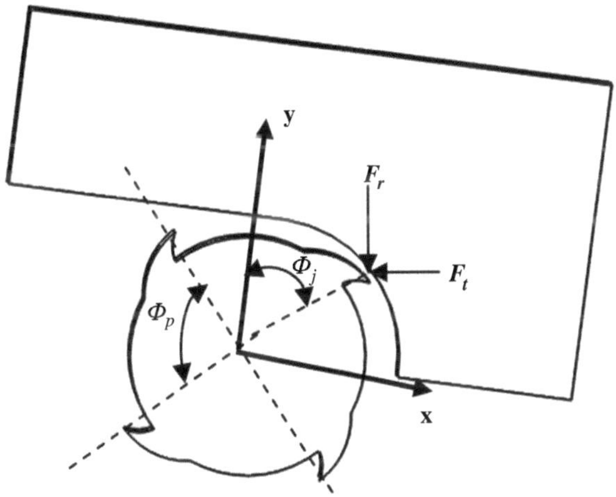

where,

$\phi_j(t)$ is the rotation angle of tooth $_j$, measured from the positive y-axis as shown in Fig. 3.

a_p presents the axial depth of cut.

K_t, K_r and K_a are the specific coefficients of cut.

The chip thickness $H(\phi_j(t))$ is the sum of a static part $H_s(\phi_j(t))$ due to the rigid motion of the cutter and a dynamic part $H_d(\phi_j(t))$ caused by the tool vibrations at the present and previous tooth periods. There are respectively expressed as follow (Faassen et al. 2003):

$$\begin{cases} H_s(\phi_j(t)) = f_z \sin(\phi_j(t)) \\ H_d(\phi_j(t)) = (u_x(t) - u_x(t-\tau)) \sin(\phi_j(t)) - (u_y(t) - u_y(t-\tau)) \cos(\phi_j(t)) \end{cases} \tag{8}$$

where,
$u_x(t)$, $u_y(t)$ represent deflections of the tool-tip at the present time,
$u_x(t-\tau)$, $u_y(t-\tau)$ are deflections of the tool-tip at the previous time,
τ is the tooth passing period time, it is defined as $\tau = \frac{60}{NZ}$

N, Z are respectively the spindle speed and the teeth number.
The rotation angle is the following:

$$\phi_j(t) = \Omega t + j\Phi_p,\ j = 0, 1, \ldots, Z-1 \tag{9}$$

Ω and Φ_p are the angular velocity and the angle between two subsequent teeth respectively, it is expressed as $\Phi_p = \frac{2\pi}{Z}$.

Table 1 Cutting parameters

Parameters	Value
Feed per tooth f_z	0.16 mm
Z	3
K_t	644 N/mm^2
K_r	0.38 N/mm^2
K_a	0.25 N/mm^2

4 Equation of Motion

The general equation of motion is obtained by applying the Lagrange's formalism to the kinetic and the potential energies. It is written as follow:

$$[M]\{\ddot{Q}\} + ([G]+[D]+[C_b(t)])\{\dot{Q}\} + ([K]+[K_b(t)])\{Q\} = \{F_{c(x,y,z)}(t,\{Q\})\} \quad (10)$$

$([M], [G], [D], [K])$ are the mass matrix, gyroscopic terms, damping matrix and the stiffness matrix respectively.

The total displacement vector composed of both elastic and rigid deformations is:

$$\{Q\} = \left[U_1, V_1, W_1, \theta_{x_1}, \theta_{y_1}, \theta_{z_1}, \ldots, U_i, V_i, W_i, \theta_{xi}, \theta_{y_i}, \theta_{z_i}, X_A, Y_A, Z_A, \alpha_x, \alpha_y, \alpha_z\right]^T$$

$$[C_b(t)] = \begin{bmatrix} 0 & . & . & 0 & . & . & 0 \\ . & . & . & . & . & . & 0 \\ . & C_{xx} & C_{xy} & . & . & . & . \\ 0 & C_{yx} & C_{yy} & . & . & . & . \\ . & .0 & . & 0 & C_{xx} & C_{xy} & . \\ . & . & . & . & C_{yx} & C_{yy} & 0 \\ 0 & . & . & . & 0 & . & 0 \end{bmatrix} : \text{variable damping coefficients} \quad (11)$$

$$[K_b(t)] = \begin{bmatrix} 0 & . & . & 0 & . & . & 0 \\ . & . & . & . & . & . & 0 \\ . & K_{xx} & K_{xy} & . & . & . & . \\ 0 & K_{yx} & K_{yy} & . & . & . & . \\ . & . & . & 0 & K_{xx} & K_{xy} & . \\ . & . & . & . & K_{yx} & K_{yy} & 0 \\ 0 & . & . & . & 0 & . & 0 \end{bmatrix} : \text{variable stiffness coefficients} \quad (12)$$

$F_{c(x,y,z)}(t,\{Q\})$ is the cutting force's vector.

Fig. 4 Stiffness coefficients variation. **a** K_{xx}, **b** K_{xy}, **c** K_{yy}, **d** K_{yx}

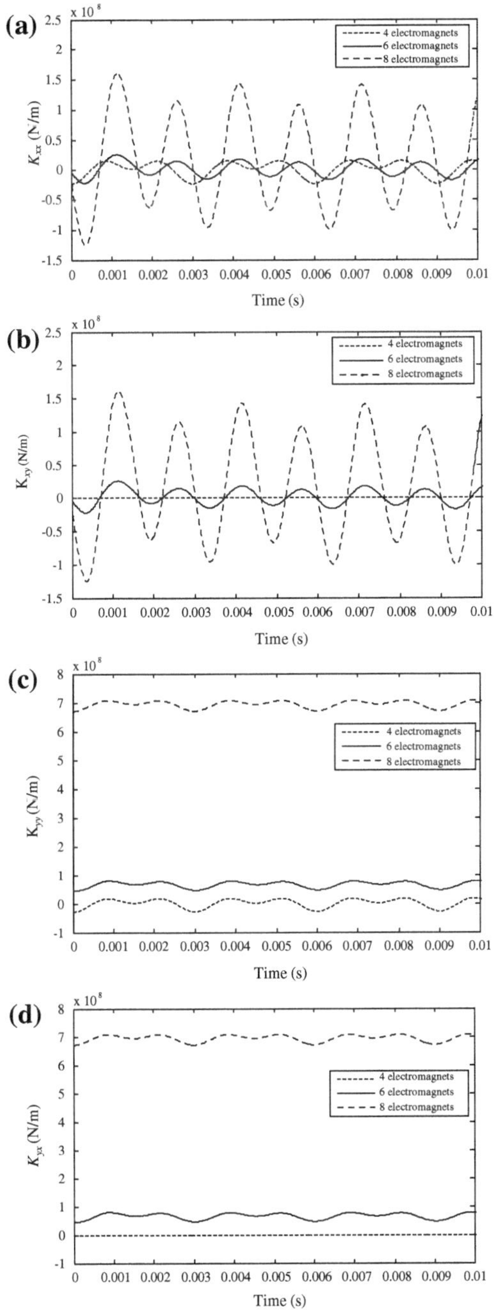

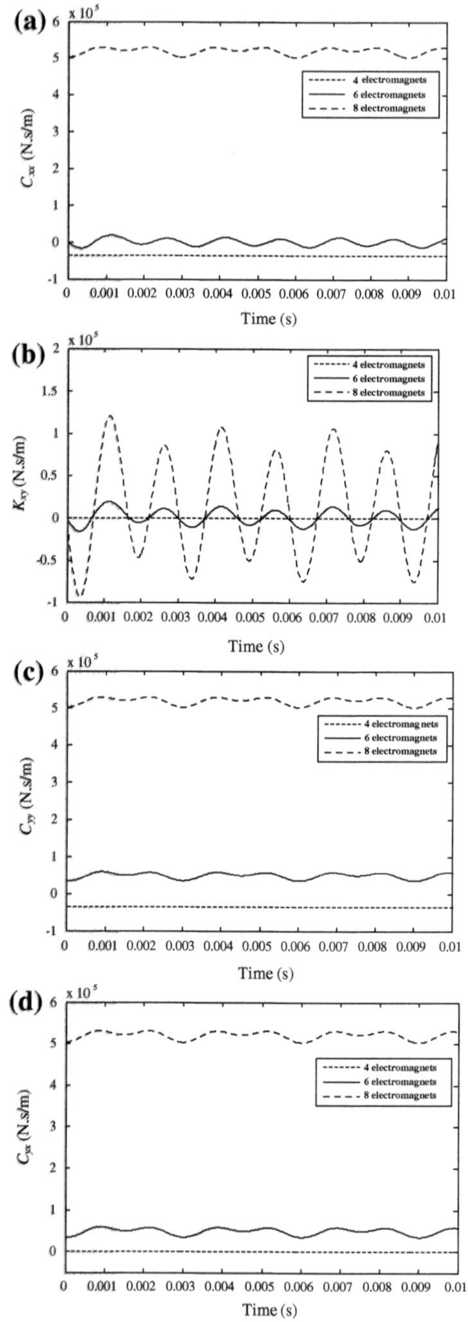

Fig. 5 Damping coefficients variation. **a** C_{xx}, **b** C_{xy}, **c** C_{yy}, **d** C_{yx}

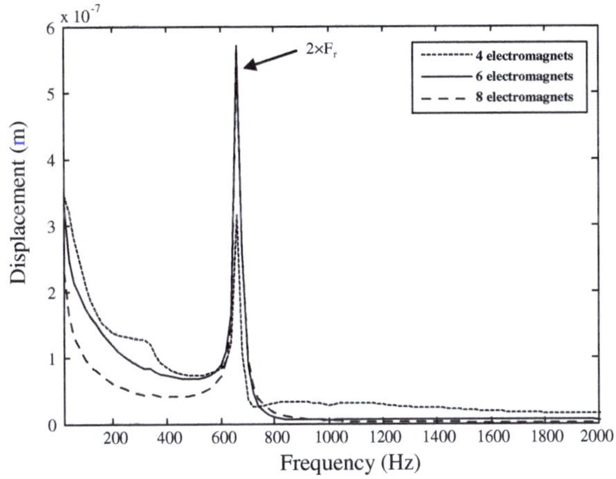

Fig. 6 Frequency spectrum of the tool tip

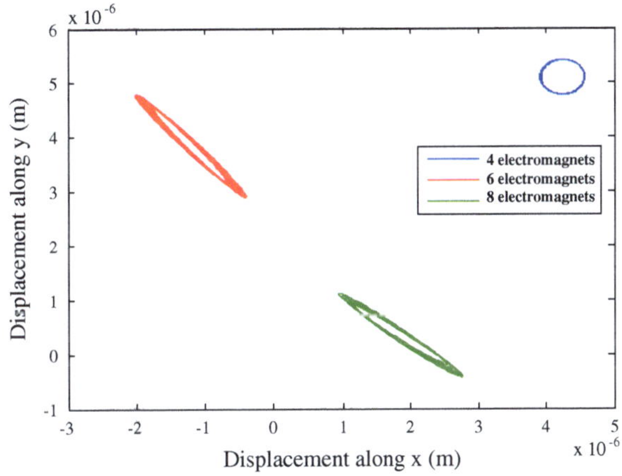

Fig. 7 Orbit of the tool tip center

5 Results and Discussion

The general dynamic equation is solved by the method of resolution of Newmark coupled with Newton Raphson. The spindle speed is assumed to be 20,000 rpm. Cutting parameters used are listed in Table 1.

Figures 4 and 5 show the time domain plot of stiffness and damping coefficients in x- and y- directions of the bottom AMB, for four, six and eight electromagnets respectively. It appears that all the components increase with the increase of the electromagnet number. Also, dynamic coefficients remain more important in the case of AMB with eight magnets. In fact, an increase in the electromagnet number leads to increase the resultant electromagnet force in each direction. Consequently, dynamic coefficient rise in order to absorb vibrations.

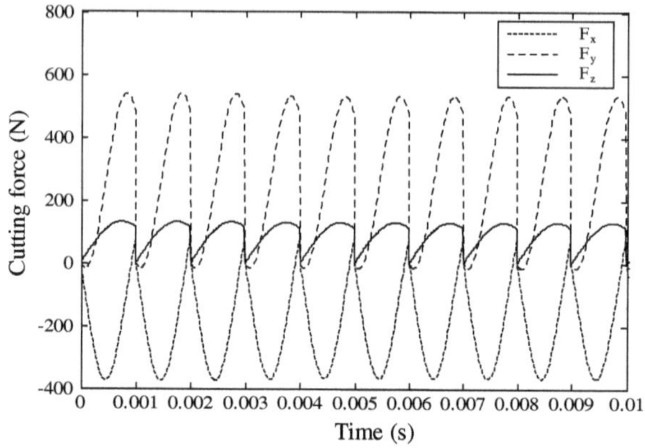

Fig. 8 Cutting force in x-, y- and z- directions for a three flute cutter and AMB with eight electromagnets

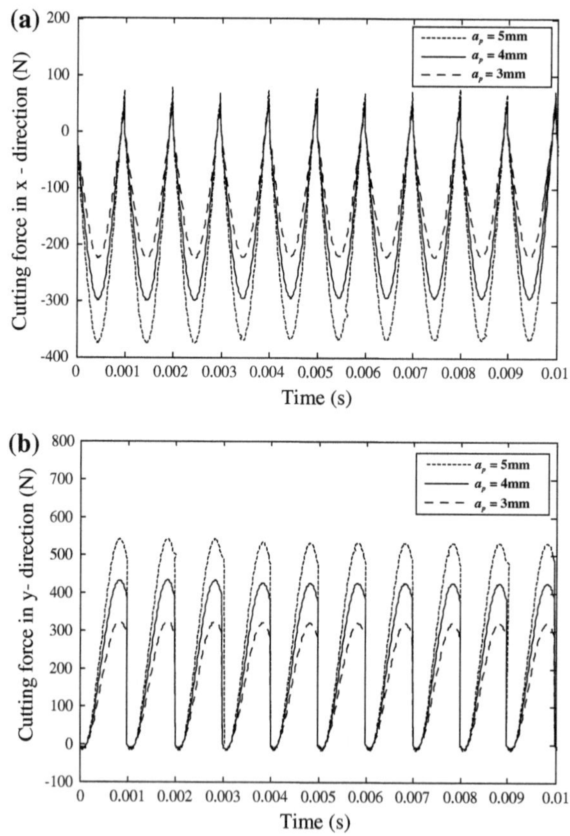

Fig. 9 Cutting force in x- and y- direction for three flute cutter (a_p: 3, 4 and 5 mm) and AMB with eight electromagnets

The coupling dynamic coefficients of stiffness and damping respectively (K_{xy}, K_{yx}) and (C_{xy}, C_{yx}) for the case of four magnet are null.

Figure 6 shows the FFT diagram for x-responses of the tool tip with two flutes at a spindle speed 20,000 rpm. It is noticed that only one major pick dominates the tool-tip response corresponding to the frequency of 2 × F_r (666.66 Hz) and which also occurs with the cutting force frequency.

Displacements of the tool tip in y- direction as function of those in x- directions for four, six and eight electromagnets are presented in Fig. 7. It can be seen that orbits have an elliptical form due to the flexibility of bearing.

Figure 8 presents the resultant cutting force for a three flutes cutter at a spindle speed 20,000 rpm, a feed rate of 0.16 mm and where eight electromagnets are used. It can be noted that cutting force distribution is continuous and constant as the cutter is in contact with the matter.

Figure 9 presents variation of the cutting force in x- and y- directions for different values of axial depth of cut a_p: 3, 4 and 5 mm and eight electromagnets. It can be seen that cutting force increase when this parameter increase. In fact, from Eq. (7) it can be noted that cutting force is proportional to the axial depth of cut.

6 Conclusion

In this chapter, an AMB spindle dynamic behavior for peripheral milling process is presented. Rigid and elastic motions are taken into account. AMB are modeled by non-linear dynamic coefficients. In fact, stiffness and damping coefficients increase with the increase of electromagnet number. Also, fluctuations at the tool tip are decreased when using more than four electromagnets. The predicted cutting force with a three teeth cutter and using eight electromagnets are continuous and increase with the increase of the axial depth of cut.

References

Auchet S, Chevrier P, Laccour M, Lipinski P (2004) A new method of cutting force measurement based on command voltages of active electro-magnetic bearing. Int J Mach Tool Manuf 44:1441–1449

Belhadj Messaoud N, Bouaziz S, Maatar M, Fakhfakh T, Haddar M (2011) Dynamic behavior of active magnetic bearing in presence of angular misalignment defect. Int J Appl Mech 3:1–15

Bouaziz S, Belhadj Messaoud N, Mataar M, Fakhfakh T, Haddar M (2011) A theoretical model for analyzing the dynamic behaviour of spatial misaligned rotor with active magnetic bearings. Mechatronics 21:899–907

Bouaziz S, Belhadj Messaoud N, Choley JY, Mataar M, Haddar M (2013) Transient response of a rotor-AMBs system connected by a flexible. Mechatronics 23:573–580

Budak E (2006a) Analytical models for high performance milling. Part I: cutting forces, structural deformations and tolerance integrity. Int J Mach Tools Manuf 46:1478–1488

Budak E (2006b) Analytical models for high performance milling. Part II: process, dynamics and stability. Int J Mach Tools Manuf 46:1478–1488

Faassen RPH, van de Wouw N, Oosterling JAJ, Nijmeijer H (2003) Prediction of regenerative chatter by modeling and analysis of high-speed milling. Int J Mach Tools Manuf 43:1437–1446

Gourc E, Seguy S, Arnaud L (2011) Chatter milling modeling of magnetic bearing spindle in high speed domain. J Mech Tools Manuf 51:928–936

Hentati T, Bouaziz A, Bouaziz S, Cholley Jy, Haddar M (2013) Dynamic behavior of active magnetic bearings spindle in high-speed domain. Int J Mechatron Manuf Syst 6:474–492

Kimman MH, Langen HH, Munning Schmidt RH (2010) A miniature milling spindle with active mahnetic bearings. Mecatronics 20:224–235

Klocke F, Qwito F, Arntz K (2009) A study of the influence of cutting parameters on micro milling of steel with Cubic Boron Nitride (CBN) tools. Micromachining and Micro fabrication Process Technology XIV 7204

Knospe CR (2007) Active magnetic bearing for machining application. Control Eng Pract 15:307–313

Lai WH (2000) Modeling of cutting forces in end milling operations. J Sci Eng. Tamkang, 3:15–22

Generalised Polynomial Chaos for the Dynamic Analysis of Spur Gear System Taken into Account Uncertainty

Manel Tounsi, Moez Beyaoui, Kamel Abboudi, Lassaad Walha and Mohamed Haddar

Abstract This chapter reviews the available literature of method taken into account uncertainty on the aerodynamics of wind turbines in the analysis of the dynamic behaviour. The focus of this work is however on the numerical modelling of the aerodynamic torque as an input taken into account uncertainties. The purpose of this overview is to include the Chaos polynomial method to take into account uncertainties according to the power coefficient of the aerodynamic torque of a two stage spur gear system with 12 DOF.

Keywords Spur gear · Aerodynamic · Uncertainty

1 Introduction

A major importance is according to the gear system because it is considering as the best solution to transmit rotational motions according to the various advantages that take such as efficiency, reliability, precision… Also, a big attention is being

M. Tounsi (✉) · M. Beyaoui · K. Abboudi · L. Walha · M. Haddar
Laboratory of Mechanics, Modelling and Manufacturing,
National School of Engineers of Sfax, Sfax, Tunisia
e-mail: manelt@yahoo.fr

M. Beyaoui
e-mail: moez.beyaoui@yahoo.fr

K. Abboudi
e-mail: kamalo1982@yahoo.fr

L. Walha
e-mail: walhalassaad@yahoo.fr

M. Haddar
e-mail: mohamed.haddar@enis.rnu.tn

M. S. Abbes et al. (eds.), *Mechatronic Systems: Theory and Applications*,
Lecture Notes in Mechanical Engineering, DOI: 10.1007/978-3-319-07170-1_11,

paid to wind turbine all over the world. So a wind turbine is a machine for converting the kinetic energy in wind into mechanical energy.

Wind turbine is a growing source of alternative energy nowadays, but it presents numerous problems such as the turbine dynamics are nonlinear and contain states that are difficult or practically not possible to measure. So it is important to introduce uncertainties in dynamic modeling.

According to the literature, there are many methods to take into account uncertainties such as Monte Carlo (Rubinstein 1981), Chaos Polynomial (Crestaux et al. 2009) (Wiener 1938), Taguchi method (Tanga and Jacques Périaux 2012).

The increased speed mechanism of the wind turbine studied is a two-stage gear system with an aerodynamic input torque. Numerous recent studies focus on the input torque as a constant variable. Or in practice, the aerodynamic torque is a random parameter since it presents dispersions related to the arbitrary behaviour of the wind… Therefore, it is necessary to take account of the above mentioned uncertainty in the dynamic response in order to predict the robustness of the analysis.

2 Dynamic Modelling

Nonlinear dynamic model of the two-stage gears is devoted in this work. The system is composed of two trains of gearings so there is three blocks as it is presents in Fig. 1. Every block (j) is supported by flexible bearing.The traction-compression stiffness is k_{yj} and the bending stiffness is k_{xj}. Also a torsional stiffness $k_{\theta j}$ is admitted for every shaft j. The wheels (11) and (32) characterise respectively the input inertia and the output inertia.

The gear mesh stiffness variation $k_i(t)$ is modelled by a square wave and it can be decomposed in two components: an average component noted by k_m, and a time variant one noted by $k_v(t)$. This function presents an alternation between one and two pairs of teeth in contact. In Fig. 2, we represents the gear mesh stiffness function of the model.

We suppose that x_i and y_j are the linear displacements of the bearing. Besides, angular displacements of every wheel are noticed by θ_{ji}.

The teeth deflection $\delta_i(t)$ is projected along the line of action.

The first and second deflections are given by:

$$\delta_1(t) = (x_1 - x_2) \cdot \sin(\alpha_1) + (y_1 - y_2) \cdot \cos(\alpha_1) + rb_{12}\theta_{12} + rb_{21}\theta_{21} \qquad (1)$$

$$\delta_2(t) = (x_2 - x_3) \cdot \sin(\alpha_2) + (-y_2 + y_3) \cdot \cos(\alpha_2) + rb_{22}\theta_{22} + rb_{31}\theta_{31} \qquad (2)$$

α represents the pressure angle (generally equal to 20°) and Rb_{ji} are the bases radius of the wheels:

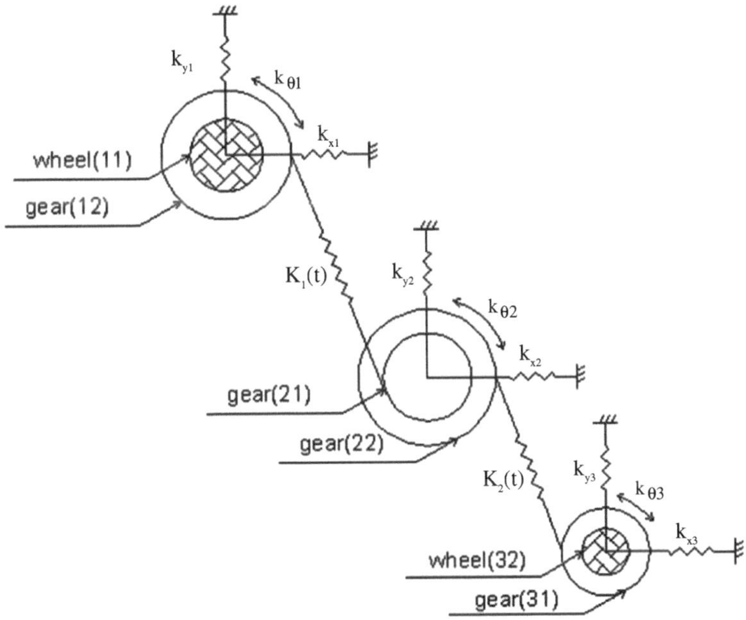

Fig. 1 Model of the two-stage gear system

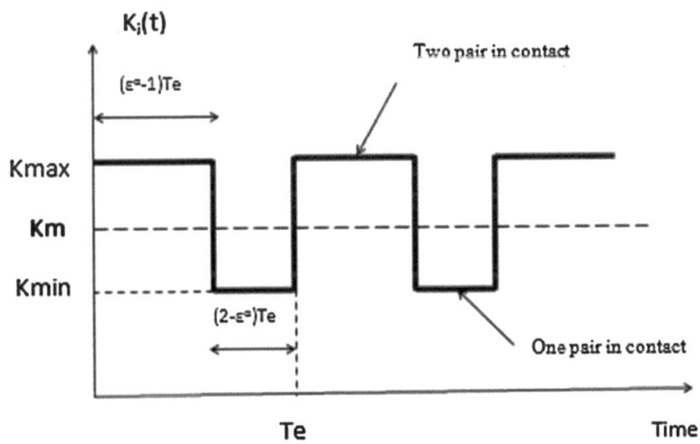

Fig. 2 Modelling of the mesh stiffness fluctuation

$$Rb_{ji} = \frac{m_{ji} Z_{ji}}{2} \quad (3)$$

3 Chaos Polynomial' Method Applied to the Motion Equation

The motion equations are presented by the following expressions:

$$\begin{cases} I_{11}\ddot{\theta}_{11} + k_{\theta 1}(\theta_{11} - \theta_{12}) = C_{a\acute{e}ro}(t) \\ I_{12}\ddot{\theta}_{12} - k_{\theta 1}(\theta_{11} - \theta_{12}) + k_1(t)Rb_{12}\delta_1(t) = 0 \\ I_{21}\ddot{\theta}_{21} + k_{\theta 2}(\theta_{21} - \theta_{22}) + k_1(t)Rb_{21}\delta_1(t) = 0 \\ I_{22}\ddot{\theta}_{22} - k_{\theta 2}(\theta_{21} - \theta_{22}) + k_2(t)Rb_{22}\delta_2(t) = 0 \\ I_{31}\ddot{\theta}_{31} + k_{\theta 3}(\theta_{31} - \theta_{32}) + k_2(t)Rb_{31}\delta_2(t) = 0 \\ I_{32}\ddot{\theta}_{32} - k_{\theta 3}(\theta_{31} - \theta_{32}) = -C_r(t) \\ m_1\ddot{x}_1 + k_{x1}x_1 + k_1(t)\delta_1(t)\sin(\alpha_1) = 0 \\ m_1\ddot{y}_1 + k_{y1}y_1 + k_1(t)\delta_1(t)\cos(\alpha_1) = 0 \\ m_2\ddot{x}_2 + k_{x2}x_2 - k_1(t)\delta_1(t)\sin(\alpha_1) - k_2(t)\delta_2(t)\sin(\alpha_2) = 0 \\ m_2\ddot{y}_2 + k_{y2}y_2 - k_1(t)\delta_1(t)\cos(\alpha_1) - k_2(t)\delta_2(t)\cos(\alpha_2) = 0 \\ m_3\ddot{x}_3 + k_{x3}x_3 + k_2(t)\delta_2(t)\sin(\alpha_2) = 0 \\ m_3\ddot{y}_3 + k_{y3}y_3 + k_2(t)\delta_2(t)\cos(\alpha_2) = 0 \end{cases} \tag{4}$$

The aerodynamic torque is given by this expression according to Lei et al. (2013):

$$C_{aero} = \rho_{air}AR^3\Omega^2C_p \tag{5}$$

With ρ_{air} represents the air density, A and R are respectively the area and the radius of the rotor, Ω is the angular velocity and finally C_p is the power coefficient.

The power coefficient is highlighted modelled by an uncertain interval $[a\ b] = [0.35\ 0.45]$ and it reacts according to a uniform distribution law:

$$C_p = \frac{b+a}{2} + \frac{b-a}{2}\xi \tag{6}$$

ξ is distributed uniformly in the interval $[-1\ 1]$.

The receiving torque is computed as follow:

$$C_r = \frac{Caero}{gear\_ratio}, \quad gear\_ratio = \frac{Z_{21}Z_{22}}{Z_{21}Z_{31}} \tag{7}$$

The Chaos Polynomial's method require in the first step to represent the system in the state space, so the system is written as the following usual matrix form:

$$\dot{z}(t) = Az(t) + f(z(t), Cp) \tag{8}$$

$$z(t) = \{\theta_{11} \quad \dot{\theta}_{11} \quad \theta_{12} \quad \dot{\theta}_{12} \quad \theta_{21} \quad \dot{\theta}_{21} \quad \theta_{22} \quad \dot{\theta}_{22} \quad \theta_{31} \quad \dot{\theta}_{31} \quad \theta_{32} \quad \dot{\theta}_{32} \\ x_1 \quad \dot{x}_1 \quad x_2 \quad \dot{x}_2 \quad x_3 \quad \dot{x}_3 \quad y_1 \quad \dot{y}_1 \quad y_2 \quad \dot{y}_2 \quad y_3 \quad \dot{y}_3\}^T \tag{9}$$

Second, we should represent any variable of the system in this form:

$$z_i = \sum_{j=0}^{P} \bar{z}_{i,j} L_j \tag{10}$$

The recurrence relation of the polynomial Legendre is given by this expression:

$$\begin{cases} (m+1)L_{n+1}(x) = (2m+1)xL_m(x) - mL_m - 1(x) \\ L_0(x) = 1,\ L_1(x) = x \end{cases} \tag{11}$$

The problem is to determine the modal coefficient noted by $\bar{z}_{i,j}$, so a Galerkin projection allows generating nonlinear deterministic differential equations system.

4 Dynamic Response

The parameters of simulation of the two-stage gear system are presented in Table 1.

The simulation of the system of differential equations is made using the ODE45 solver of Matlab in order to obtain the dynamic behaviour of the two stage spur gear system with consideration of uncertainties in the aerodynamic input torque and more precisely in his power coefficient.

Aerodynamic torque is presented in Fig. 3, it is a sinusoidal signal fluctuating between zero and 25 Nm. the interval of fluctuation is in relation with the boundary condition.

Figure 4 represents the resultant of the displacements of the first bearing. These displacements are defined following two directions x and y. The amplitudes of displacement are of the order of 10^{-5} m and fluctuate in the positive part. This result is the same on the third bearing. We plotted on Fig. 5 the resultant of the displacement of the third bearing.

These results are compared by the system without consideration of uncertainties according to the aerodynamic input torque. In Figs. 6 and 7 we represent the resultant of displacement of the first and third bearing respectively. Figures shown that the signal fluctuate with an amplitude of 10^{-5} m in both x and y-direction.

Table 1 Parameters of simulation

Description	Symbol	Value
Material	42CrMo4	–
Gear density	ρ	7860
Generator moment of inertia	J_{gener}	20
Wind turbine moment of inertia	J_{turb}	2895
Turbine rated speed	Ω	13
Rotor diameter	D	12
Stiffness to bending	k_{xj}	7×10^8
Stiffness to traction—compression	K_{yj}	6×10^8
Average mesh stiffness	k_m	2×10^8
Torsional stiffness of the shaft	$k_{\theta j}$	5×10^6
Width of teeth	b	0.1
Number of teeth	Z(12)	72
	Z(21)	18
	Z(22)	54
	Z(31)	18
Module of teeth	m	0.016
The pressure angle	α	20
Contact ratio	$\varepsilon_{\alpha 1}$	1.67
	$\varepsilon_{\alpha 2}$	1.64

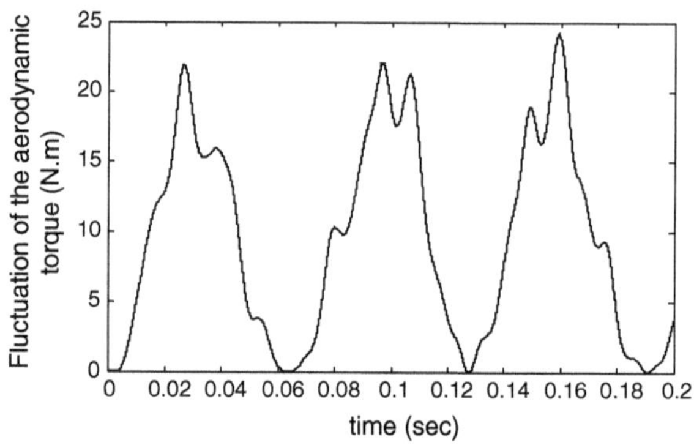

Fig. 3 Fluctuation of the aerodynamic torque

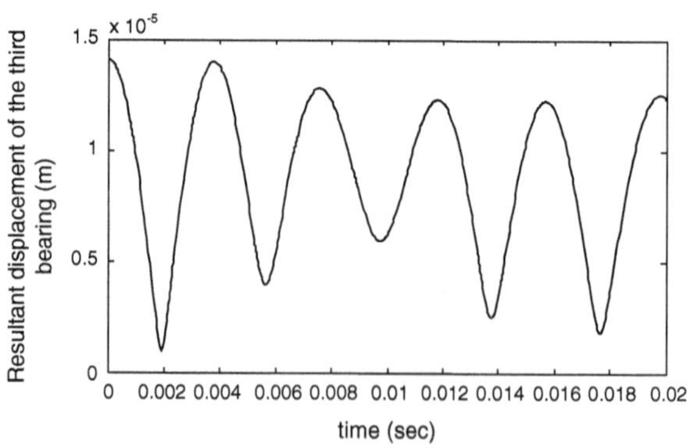

Fig. 4 Temporal features of the dynamic components of the first bearing

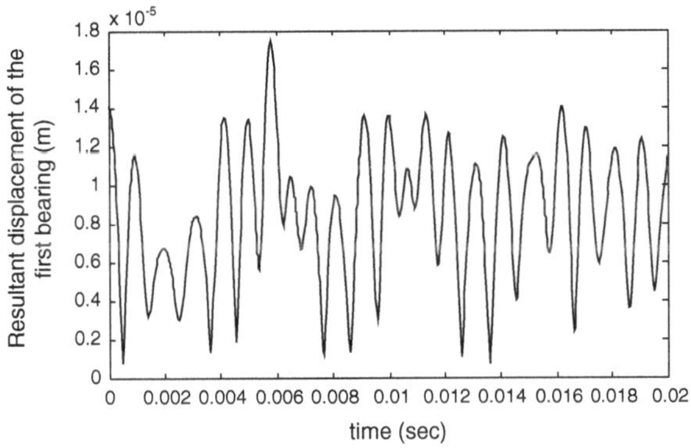

Fig. 5 Temporal features of the dynamic components of the third bearing

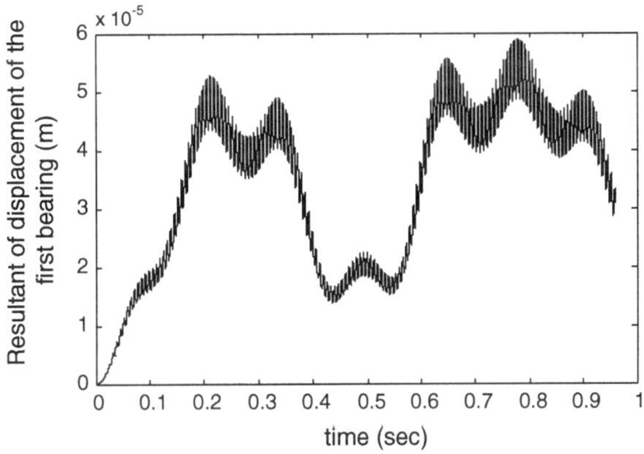

Fig. 6 Resultant of displacement of the first bearing without uncertainties

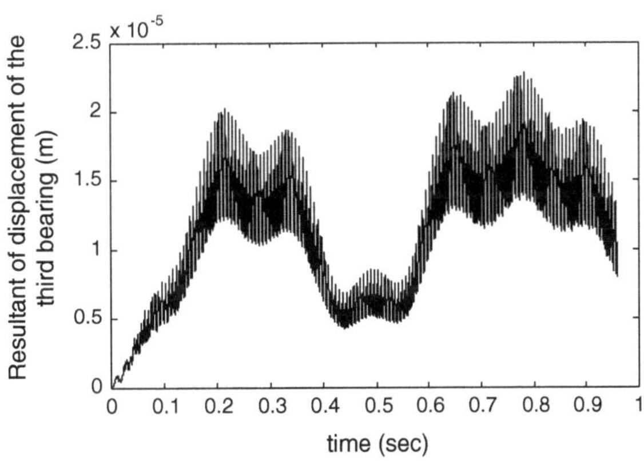

Fig. 7 Resultant of displacement of the third bearing without uncertainties

5 Conclusion

In order to ensure robustness in the analysis of the dynamic response of a mechanical system, it is very important to take into account uncertainties in the phase of design modelling. In this chapter, we introduce uncertainty in the power coefficient of an aerodynamic input torque of a two-stage gear system with 12 degrees of freedom and here we represents the novelty of this work because in recent studies, many authors consider this aerodynamic torque as constant input or these hypothesis takes us away from the reality of things. So to obtain deterministic system, we applied the Chaos Polynomial method to take into account uncertainties of the power coefficient which is considered by an uncertain interval.

References

Crestaux T, Le Maitre O, Martinez JM (2009) Polynomial chaos expansion for sensitivity analysis. Reliab Eng Syst Saf 94:1161–1172

Lei Y, Bai Y, Xu Z, Gao Q et al (2013) An experimental investigation on aerodynamic performance of a coaxial rotor system with different rotor spacing and wind speed. Exp Therm Fluid Sci 44:779–785

Rubinstein RY (1981) Simulation and the monte carlo method. Wiley, NewYork

Tanga Z, Jacques Périaux J (2012) Uncertainty based robust optimization method for drag minimization problems in aerodynamics. Comput Methods Appl Mech Eng 217–220:12–24

Wiener N (1938) The homogeneous chaos. Am Jo Math 60(4):897–936

Modelling and Simulation of the Doctors' Availability in Emergency Department with SIMIO Software. Case of Study: Bichat-Claude Bernard Hospital

Mahmoud Masmoudi, Patrice Leclaire, Vincent Cheutet and Enrique Casalino

Abstract Emergency Departments (ED) require an appropriate allocation of human and material resources in order to increase their effectiveness and efficiency and reduce as much as possible the patients' waiting time. This chapter describes step by step the patient's stay process at the ED. The SIMIO software was used for the modelling and simulation of this process. This chapter also presents a novel method of modelling doctors' availability in the emergency department taking into account their number and availability in trauma and medicine areas. The findings from the different simulated scenarios show that modifying the doctors' number can have a strong effect on the patients' length of stay and the number of exited patients from the service. This work was based on the ED at Bichat Claude Bernard Hospital in Paris, France.

Keywords Modelling · Simulation · Emergency department · SIMIO software

M. Masmoudi (✉)
Mechanics, Modelling and Manufacturing Laboratory (LA2MP), Mechanical Engineering Department, National School of Engineers of Sfax (ENIS), BP 1173, 3038 Sfax, Tunisia
e-mail: masmoudi.miha@gmail.com

P. Leclaire · V. Cheutet
Laboratory of Mechanical Systems and Materials (LISMMA), Higher Institute of Mechanics of Paris (SUPMECA), 3 rue Fernand Hainaut, 93407 Saint-Ouen, France
e-mail: patrice.leclaire@supmeca.fr

V. Cheutet
e-mail: vincent.cheutet@supmeca.fr

E. Casalino
Emergency Department, Bichat-Claude Bernard Hospital, 46 Rue Henri Huchard, 75018 Paris, France
e-mail: enrique.casalino@bch.aphp.fr

M. S. Abbes et al. (eds.), *Mechatronic Systems: Theory and Applications*,
Lecture Notes in Mechanical Engineering, DOI: 10.1007/978-3-319-07170-1_12,

1 Introduction

In recent years, applied researchers have become increasingly interested in modelling patients' flow in hospital in order to optimize its services.

Most of the time, the management of these flows is a complicated task, because of the increasing number of patients at hospital. For an Emergency Department, the patients come according to a random inter-arrival time and therefore the queue can be expanded during the day, and from one day to another. Thus, the control of the patient's pathway is considered nowadays as a major issue. According to Jlassi (2011), the main objective of an emergency department is to ensure patients' support rapidly and qualitatively with an efficient planning of the hospital's resources.

Recent studies have begun to explore various tools and methods in the medical context. Wang et al. (2009) suggested ARIS and Arena to model and simulate an emergency service in Lyon, France. Su and Shih (2003) focused on modelling 23 EMS hospitals in Taipei, Taiwan to perform two-tier rescues using computer simulations. Ahmed and Amagoh (2008) use the AWESIM simulation model to analyse human resource requirements in a hospital.

Nevertheless, using the SIMIO software, available at SIMIO (2007), in the context of modelling and simulating patients' flow, taking into account the doctors' availability in the Emergency Department, is as yet an unexplored territory.

The aim of the present chapter is to give a suggestion for modelling and evaluating the impact of the change of the doctors' number on the patient's length of stay (LOS).

The remainder of this chapter is divided into three sections. Section 2 describes the patient's stay process: the steps and actions made by the patient from his ED entry to exiting in a chronological order. Section 3 provides details about modelling Bichat ED, the considered hypotheses and the different used modelling concepts. The last section is an evaluation of the variation of the physicians' number who are present at the ED.

2 The Patient Stay Process

Every year, more than 72,000 patients visit the Bichat emergency department; this number is increasing from one year to another. This emergency department receives an average of 190 patients per day with a variable inter arrival time.

The patient stay process includes many steps: Patients' arrival, Admission and registration, Waiting, Triage, Consultation, Additional analyses and Exit (see Fig. 1).

Patients can access to the ED by either one of the two possible gates depending on their arrival mode: personal vehicle or ambulance.

As soon as they arrive to the ED, either standing or in lying position, they go to the reception. A nurse is installed there to admit and register patients. She asks

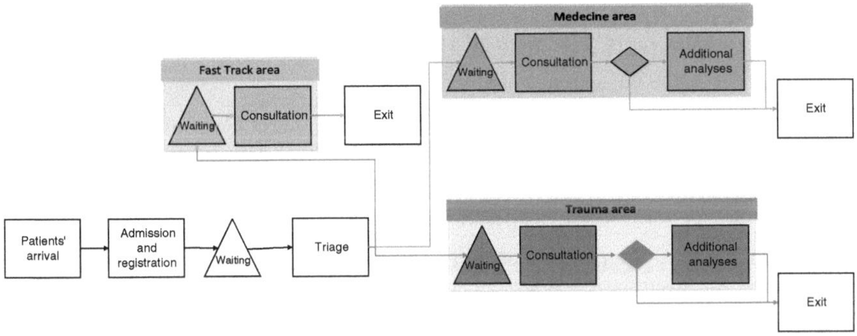

Fig. 1 Patient stay process

them about their identity and social security number and checks whether they already benefited from the hospital's services.

Generally, the first arrived patient is the first served (FIFO) with some exceptions, such as giving priority to lying patient in a serious condition.

After the admission of the patient by a nurse, the patient is installed in one of the waiting rooms: sitting waiting room or lying waiting room. This assignment is done according to the patient's condition: standing or lying.

Then two triage nurses (TN) when available, move from their boxes to the reception to retrieve the next patient's medical chart. Each patient is installed in a triage box and asked about his medical history (earlier diseases, surgeries, allergies, current treatment, etc.) and to help the triage nurse to identify his health condition. TN decides subsequently to which area the patient should be oriented. According to the pathology type, three choices are possible:

- Fast Track area.
- Trauma area (or blue area).
- Medicine area (or red area).

The pathology type is rated from 1 to 5 according to this scale of cases gravity:

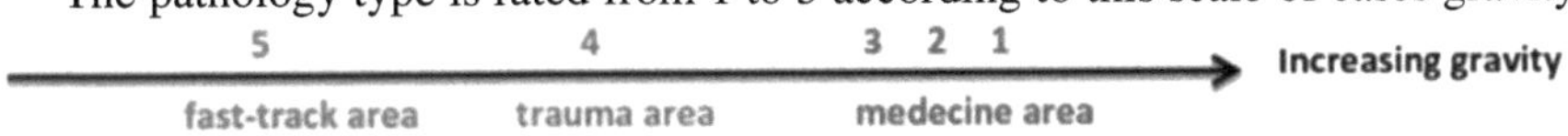

At the end of triage, the nurse installs the patient in the waiting room and re-turns to the reception to take care of another patient.

Patients with gravity level 5 return to the reception waiting area and wait to the consultation of a Fast Track doctor. Generally, these patients leave the Emergency Department quickly.

A part of standing and lying patients are oriented towards the trauma area be-fore waiting for the doctor consultation in the trauma waiting room. A nurse is responsible for installing them in the trauma boxes. Two doctors are placed into the blue area to support patients and take one of the following three decisions:

- If patients have no trauma problems and need treatments or additional analyses, they are therefore allowed to leave the emergency service.

- If patients have trauma problems and do not need further analysis, they are in this case supported by a nurse in their boxes to be treated. After treatment, patients are allowed to leave the service.
- If patients have trauma problems and are in need of additional analysis, they go to the radiology department then return to settle in the plastering box. Plastering requires the availability of a box and a doctor simultaneously. After plastering, these patients leave the ED.

The remainder of the patients are directed towards one of the two holding areas in the medicine area:

- A sitting room with infinite capacity for patients who are not in a critical state and can sit.
- A lying waiting room for lying patients.

When a medicine box is available, the first arrived patient in one of the waiting rooms is installed into before being consulted by a red area's doctor. They examine the patient and prescribe the medical review or the additional analyses needed, then take one of the following three decisions:

- A sitting room with infinite capacity for patients who are not in a critical state and can sit.
- A lying waiting room for lying patients.
- Recumbent patients who do not require any additional analyses are sent to the lying waiting room (6 places) where they await the results of the medical review and stay under nurses' observation. Following the treatment of their medical condition by a doctor, they are transferred to the hospital or to "the door service" to be hospitalized. (The door service is an area that is located next to the emergency department that hosts hospitalized patients or transferred to other hospitals).
- Patients (standing or lying) who need further analysis, leave the medicine box to do their analyses then settled back in one of the mentioned above waiting rooms:
  - Sitting Waiting room to treat their disease if they have to leave the service.
  - Lying Waiting room to put them under observation before hospitalization.

3 Modelling Doctors' Availability with SIMIO

At the ED, an emergency doctor (internal or senior) must treat and admit patients whose state of health necessitates a quick and fast support at any time of day or night. They examine the patient and make a diagnosis in a little time. That is why we find a dynamic number of physicians in Bichat hospital ED depending on time.

Table 1 Doctors number in the emergency department

Day/Time	[8:30 A.M.–2:00 P.M.]	[2:00–6:30 P.M.]	[6:30–12:00 P.M.]	[0:00–8:30 A.M.]
From monday to saturday	8	9	6	4
Sunday	4	6	6	4

Below Table 1 is a summary of the number of physicians on duty in the emergency service for each day of the week and each time slice of the day during the period of investigation.

The availability of the physician is not always equal to 100 % of their working time. In fact, the Fast track physician is able to treat 4 patients per hour, while a medicine or trauma doctor can process 1.6 patients per hour.

This is due to the complexity of their tasks; they consult patients, complete administrative tasks, review the additional analyses results and research on the patient disease before making their medical decision. We can classify these tasks into three types, each representing 1/3 of their availability in the ED.

The first part presents the patient first contact and consultation in his\her box, the second corresponds to reflection and decision-making time. The rest contains displacements inside the Emergency Department, breaks, and entering the data into the hospital information system.

To model the last 1/3, it was assumed that the doctor makes a break of 15 min every 45 min. And to randomize these events, we assigned the duration of the break to a uniform distribution with parameters (min = 10 min, max = 20 min) and the inter arrival of breaks to a uniform distribution with parameters (min = 40 min, max = 50 min).

3.1 *Modelling the Blue Area Doctors*

For modelling the availability of blue area doctors, we adopted the following hypotheses:

- A blue doctor means a physician assigned to blue area.
- 2 blue doctors are present at the Emergency Department 24 h a day and 7 days a week.
- If there is one patient in the blue box and two doctors are available, the 1st doctor supports the patient.
- The second physician is the only responsible for the plastering. To model this in SIMIO, we used 2 support servers: "consultation_doctor_b1" and "consultation_doctor_b2" followed by two decision servers « decision_doctor_b1 » and « decision_doctor_b2 » (see Fig. 2).

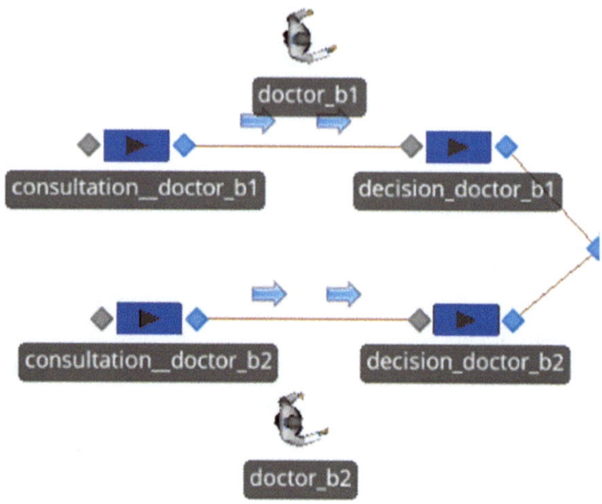

Fig. 2 Modelling blue area's doctors

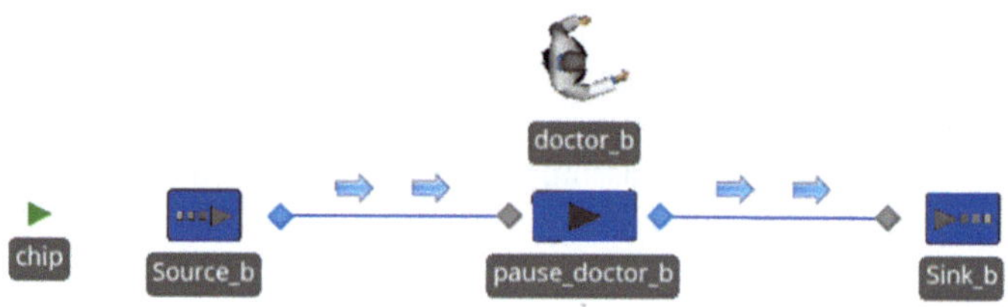

Fig. 3 Modelling a blue doctor break

Each of these servers is characterized by a fixed capacity equal to 1, a processing time equals 15 min, a blue doctor as a secondary resource and the ranking rule "FIFO".

In order to model the remaining time of the doctor, we created for each of the two blue doctors a subsystem (see Fig. 3). This subsystem is composed of a chips source that randomly creates one by one with a time of inter arrival following a uniform law with parameters (min = 40 min, max = 50 min). Then these chips move to a server called "pause_doctor_b" where they await for the passage of the "doctor_b". The duration of the pause also follows a uniform law with the parameters (min = 10 min, max = 20 min). Finally, the chips that were created exit through "sink_b".

To validate this model, we tested it for a blue doctor who is available 24 × 7 h. After a full day simulation, it was found as results:

- The number of patients entered and came out of the 'consultation_doctor_b1' server are both equal to 33 patients/day. So, we can say that the doctor consulted 33 patients in their blue boxes.

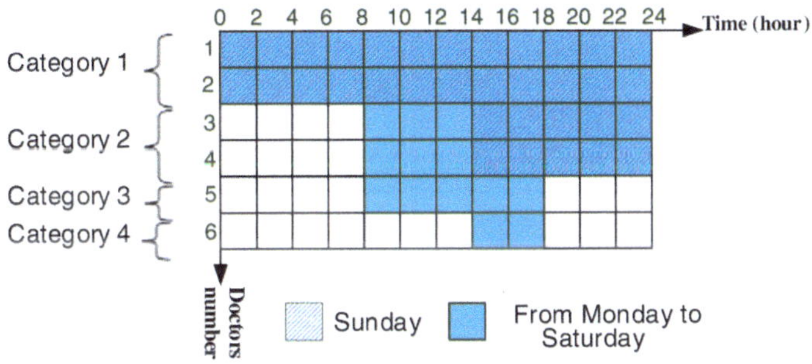

Fig. 4 Blue area doctors planning

- The number of patients entered and came out of the server 'decision_doctor_b1' are equal to 33 patients per day and 32 patients/day respectively; which means that the doctor consulted 33 patients but there is still a patient waiting in his box without decision.

4 Modelling the Red Area Doctors

The availability of doctors in the red area is different from the blue one because it has a variable number of physicians during the day.

In Fig. 4, the hatched blue color represents the doctors' availability on Sunday and the blue color for the other days. To model this constraint, we classified the doctors into 4 categories.

For that, we created 5 day patterns, indicating the start time and the end time of service for each category and 4 work schedules under SIMIO software.

Each day pattern presents a planning schedule for one day (Fig. 5). And each work schedule (Fig. 6) is a weekly planning, setting for each day the matching day patterns.

Then, the same modelling approach as for blue doctors was adapted for the red sector physicians. A red subsystem was created to model the availability of the doctors. The hypothesis that were used are as follows:

- The existence of 2 red doctors present 24 h a day and 7 days a week at the emergency service.
- When a patient is in a red box and several red doctors are available, it is the doctor who has the greatest priority value that supports the patient (doctor_r1 has the first priority and doctor_r6 has the least one).

Under SIMIO, 6 support servers "consultation_doctor_r" were used and followed by other six decision servers "dec_doctor_r" (see Fig. 7). Each of these servers is characterized by a fixed capacity equal to 1, a service time "processing

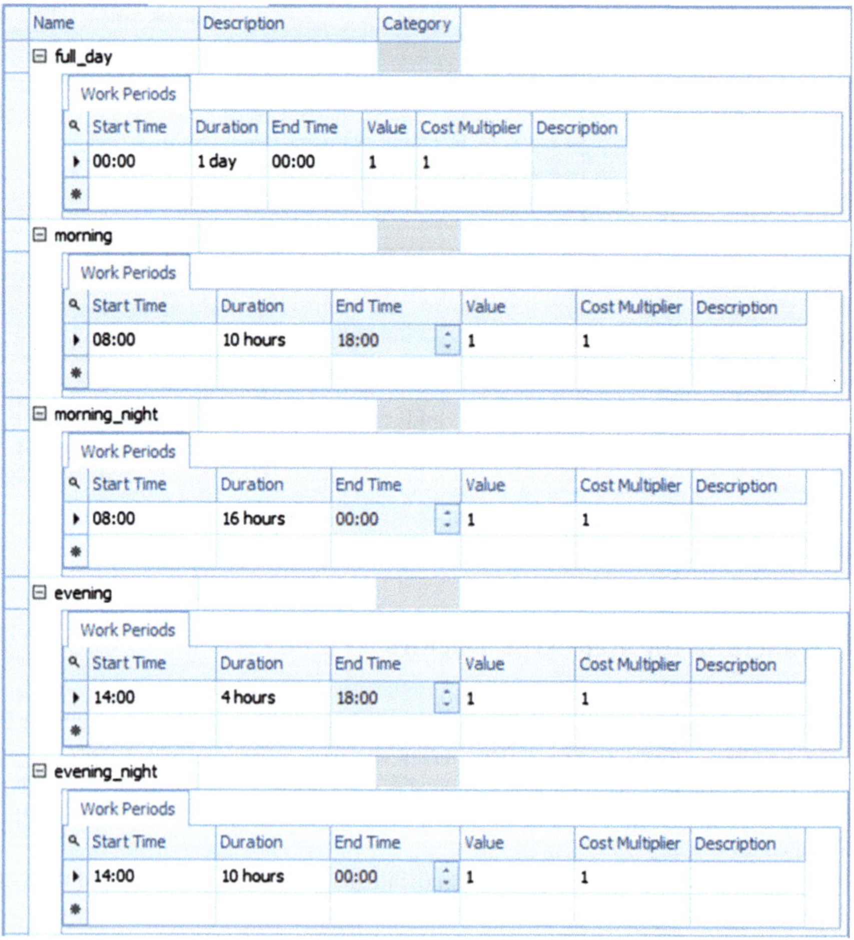

Fig. 5 Day patterns definition in SIMIO software

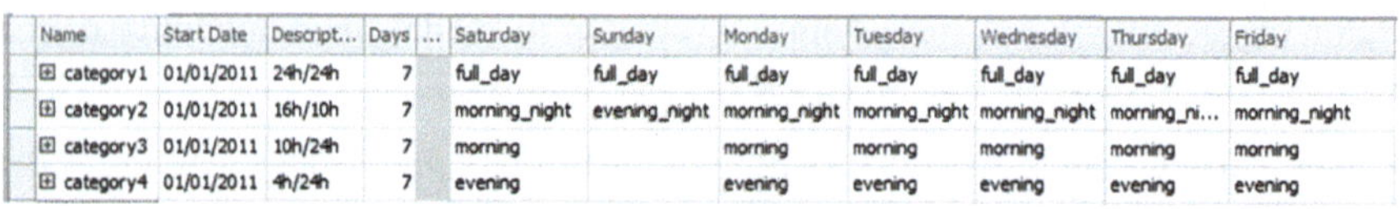

Name	Start Date	Descript...	Days	...	Saturday	Sunday	Monday	Tuesday	Wednesday	Thursday	Friday
category1	01/01/2011	24h/24h	7		full_day	full_day	full_day	full_day	full_day	full_day	full_day
category2	01/01/2011	16h/10h	7		morning_night	evening_night	morning_night	morning_night	morning_night	morning_ni...	morning_night
category3	01/01/2011	10h/24h	7		morning		morning	morning	morning	morning	morning
category4	01/01/2011	4h/24h	7		evening		evening	evening	evening	evening	evening

Fig. 6 Work schedules definition in SIMIO software

time" equal 15 min, a red doctor as a secondary resource and FIFO "Ranking Rule".

To model the red doctors availability, this subsystem is composed of source that randomly creates chips one by one with a time of inter arrival following a uniform

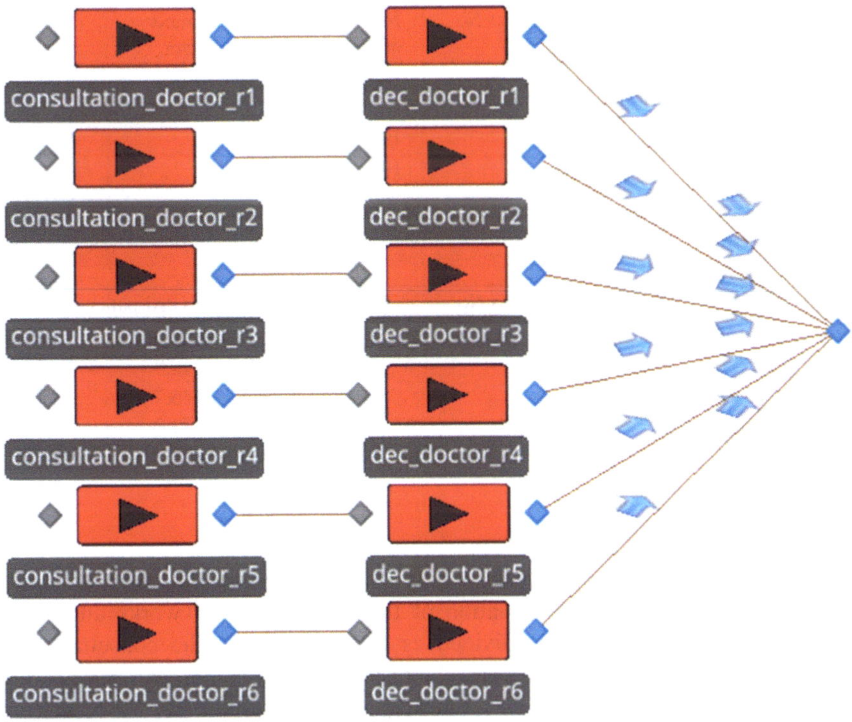

Fig. 7 Modelling red area's doctors

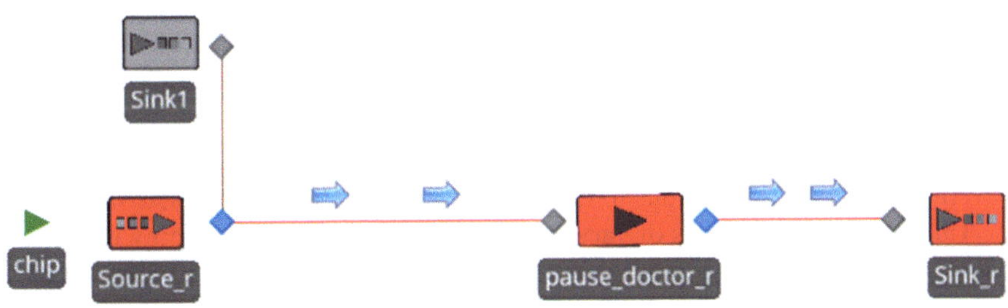

Fig. 8 Modelling a medicine doctor break

law of parameters (min = 40 min, max = 50 min). Then these tokens go through a server called "pause_doctor_r" where they await the passage of the "doctor_r". The duration of the pause follows a uniform law with parameters (min = 10 min, max = 20 min). Finally, chips exit through "sink_r".

In creating these chips for 24 h of simulation, we may exceed 33 % of the availability of a doctor belonging to certain categories which do not cover all of the 24 h. For this, we need to destroy some of them. These extra chips are rejected out from the beginning to the exit "sink1". This path is provided for the chips arriving when the doctor is not available in these periods of times (see Fig. 8).

Table 2 Red area physician's number change

Scenario No	Number of replications	Number_doctor_r1	Number_doctor_r2
1	100	1	1
2	100	2	1
3	100	1	2
4	100	2	2
5	100	3	3

Table 3 Simulation results

Scenario No	LOS average	Exited patients number
1	413,769	128,9
2	354,422	148,9
3	370,946	145,7
4	332,359	152,3
5	315,401	156

After simulation (of one week), we found a coincidence of the simulation results with reality observed situation of the ED; the doctor works during his schedule of work, his time is divided as follows: 1/3 for the consultation of patients, 1/3 for decision-making and the rest for breaks. So we can validate our model.

5 Variation in the Number of Doctors

The emergency physicians wanted to know the influence of the increase and decrease of the number of doctors in the emergency on the LOS. We conducted several simulation replications to ensure the reliability of gathered statistical data.

This experiment consists of varying the number of doctors belonging to the first category; who are always present in the red area for two days of simulation. Five scenarios were created (see Table 2).

The variable 'number_doctor_r1' refers to the first doctor of category 1 in the red area. If this value is equal to 2, it means that the number of doctors is multiplied by two. Similarly, the variable "number_doctor_r2" presents the second doctor in the first category in the red zone. For example, the first scenario shows our basic model (compliant with reality: doctor_r1 + doctor_r2) and the second scenario is to add a third red doctor who shall share with the "doctor_r1" their medical and administrative tasks.

After 100 replications of these various scenarios, we obtained the results presented in Table 3.

From these results, we note that the addition of a doctor leads to decrease of the LOS average time of a patient's case gravity 1, 2 or 3 and also to increase the number of patients exited from the ED.

In addition, from the discrepancy of results between the second and the third scenarios, we can say that the addition of a doctor_r1 is more advantageous than the addition of a doctor_r2.

6 Conclusion

We developed with SIMIO software an Emergency department model and used it to test several scenarios by simulation, changing one of its parameters each time. The main idea of this chapter is to suggest a way to model the doctors in the emergency department taking into account their availability constraints in order to get more precise results conforming to the real values.

Further short-term work will focus on issues of integrating the model into a tool for decision support for optimization problems such as the medical resource allocation. Another issue is to check the influence of the increase in the number of patients arriving at the emergency department in the coming years and to develop a methodology to account for the optimum number of doctors that the ED should have.

Acknowledgements The authors acknowledge the precious collaboration of the emergency department in Bichat-Claude-Bernard Hospital; we would like to thank all the staff, and particularly Pr. CASALINO Enrique, Emergency Department Director, for their time and effort contributed to the present research work. The authors also gratefully acknowledge the helpful comments and suggestions of the reviewers, which have improved the presentation.

References

Ahmed S, Amagoh F (2008) Modelling hospital resources with process oriented simulation, vol I, No. 1, Central Asia Business, ISSN 1991-0002

Jlassi J (2011) Amélioration de la performance par la modélisation des flux logistiques des patients dans un service d'urgence hospitalier

SIMIO (2007) Simio, process flow simulation software 2D/3D, Net Prints. http://www.simio.com. Accessed 11 March 2013

Su S, Shih C-L (2003) Modelling an emergency medical services system using computer simulation. Int J Med Inform 72:57–72

Wang T, Guinet A, Belaidi A, Besombes B (2009) Modelling and simulation of emergency services with ARIS and Arena. Case study: the emergency department of Saint Joseph and Saint Luc hospital, Production Planning & Control, vol 20, No. 6, Sept 2009, pp 484–495

FGM Shell Structures Analysis Using an Enhanced Discrete Double Directors Shell Element

Mondher Wali, Abdessalem Hajlaoui, Jamel Mars, K. El Bikri, Abdessalem Jarraya and Fakhreddine Dammak

Abstract In order to examine the accuracy of the enhanced double directors shell model for the functionally graded material (FGM) shell structures, a series of benchmark static tests are tackled using finite elements method. For implementing the discrete double directors shell model (DDDSM) within the Enhanced Strain Technique (EST), four parameters are used for enhancing the membrane strain. The vanishing of transverse shear strains on top and bottom faces is considered in a discrete form. The mechanical properties of the shell structure are assumed to vary continuously in the thickness direction by a simple power-law distribution in terms of the volume fractions of the constituents. A comparison with the exact solutions presented by several authors for shell structures indicates that the present model accurately estimates the stress-strain responses.

Keywords FGM shell structures · Enhanced strain · Third-order theory

1 Introduction

The increasing use of FGMs, as heat-shielding materials, in aerospace engineering, has clearly demonstrated the need for the development of new theories to efficiently and accurately predict the behavior of such structural components. The intrinsic in-homogeneity of these composite structures, coming from the continuous variation

M. Wali (✉) · A. Hajlaoui · J. Mars · A. Jarraya · F. Dammak
Mechanical Modelisation and Manufacturing Laboratory (LA2MP), National Engineering School of Sfax, University of Sfax, W3038 Sfax, Tunisia
e-mail: mondherwali@yahoo.fr

K. El Bikri
LaMIPI, ENSET de Rabat, Rabat-Instituts, P.O 6207 Avenue de l'Armée Royale, Madinat Al Irfane, 10100 Rabat, Morocco

M. S. Abbes et al. (eds.), *Mechatronic Systems: Theory and Applications*, Lecture Notes in Mechanical Engineering, DOI: 10.1007/978-3-319-07170-1_13,

of properties in one or more directions, makes the classical theories of shells inadequate.

The study of FGM structures using classical theory, based on the Kirchhoff hypothesis, is lack of precision. The inaccuracy is due to neglecting the effects of transverse shear and normal strains of the structure. The inspiration and guidelines for the subsequent attempts have stemmed from Reddy (2000) and Zenkour (2006)'s works where the exact Navier solutions for bending analysis of a simply supported FGM plates were given.

In recent years, the various first-order and higher-order shear deformation theories were proposed to analyze the FGM plates. The models based on the first-order shear deformation theory (FSDT) are very often used owing to their simplicity in analysis and programming. It requires however a convenient value of the shear correction factor (Nguyen et al. 2008). In practice, this coefficient has been assumed to be given by 5/6 as for homogeneous plates. This value is a priori no longer appropriate for functionally graded material analysis due to the position dependence of elastic properties. Nguyen et al. (2008) identified the shear correction coefficients for the FSDT models made of functionally graded materials. Applications are presented for a simply supported plate and for a sandwich panel which is clamped at both ends. The influence of this factor on the static response is then presented. Singha et al. (2011) employed a four-node high precision plate bending element based on the exact neutral surface position and the first- order shear deformation theory to analyze functionally graded plates subjected to sinusoidal or uniformly distributed lateral loads. The shear correction factor is calculated from the energy equivalence principle. Castellazzi et al. (2013) presented the nodal integrated plate element (NIPE) formulation for the analysis of functionally graded plates based on the first-order shear deformation theory. The strain-displacement operators are derived via nodal integration, for linear triangles and quadrilateral elements.

To avoid this difficulty, several authors proposed the higher-order shear deformation theory (HSDT) and applied it to FGM. Reddy (2000) presented a general formulation for FGMs using the third-order shear deformation plate theory and developed the associated finite element model that accounts for the thermo-mechanical coupling and geometric non-linearity. Xiao-Hong et al. (2002) presented a high order theory to model the electromechanical behavior of functionally graded piezoelectric generic shells. The generalized Hamilton's principle, which incorporates different electric boundary conditions as well as mechanical boundary conditions, is utilized to obtain the governing equations of motion. Ferreira et al. (2007) studied the static deformations of functionally graded plates using the radial basis function collocation method and a higher-order shear deformation theory. They select the shape parameter in the radial basis functions by an optimization procedure based on the cross validation technique. Carrera et al. (2011) evaluated the effect of thickness stretching in plate/shell structures made by materials which are functionally graded (FGM) in the thickness directions. Xiang and Kang (2013) used an n-order shear deformation theory and a meshless global collocation method based on the thin plate spline radial basis function to analyze the static characteristics of functionally graded plates under sinusoidal load.

The aim of this chapter is to investigate the accuracy of the DDDSM formulation for static analysis of functionally graded shell structures. In general this approach, called also a multi-director shell theory, was used by several authors such as Başar et al. (2000), Brank and Carrera (2000), Brank et al. (2002), Brank (2005) to study the behavior of the multi-layered structures including finite-rotation theory and studying the-thickness stretching effect. In the DDDSM formulation, the vanishing of transverse shear strains on top and bottom faces is considered in a discrete form. This formulation is enhanced by using the Enhanced Strain Technique (EST). The EST, introduced by Simo and Rifai (1990), consists in augmenting the space of discrete strains with local functions, which may not derive from admissible displacements. A suitable choice of these additional modes can improve the numerical performance of the shell elements. In our work, the compatible in-plane strain is enhanced by using four parameters to improve the membrane behavior of the bilinear shell element. This chapter treated in validation tests only linear applications for FGM shell structures. It is noticed that the present DDDSM formulation can predict accurately the dimensionless displacements and bending stresses of simply supported FGM shell structures.

2 Double Directors Shell Model

In this section, the geometry and kinematic of nonlinear double directors shell model are described. The fixed spatial coordinate system is defined by a triad ($\boldsymbol{E}_{\mathrm{i}}$), i = 1, 2, 3. The reference surface of the shell is assumed to be smooth, continuous and differentiable. Initial and current configurations of the shell, are denoted, respectively, by C_0 and C_{t}. Variables associated with the undeformed state C_0 will be denoted by upper-case letters and by a lower-case letter when referred to the deformed configuration C_{t}. Also, the vectors notation will be denoted by bold face letters.

2.1 Double Directors Shell Kinematic Hypothesis

Parameterizations, which define material points of the shell, are carried out in terms of curvilinear coordinates $\boldsymbol{\xi} = \left(\xi^1, \xi^2, \xi^3 = z\right)$. The position vectors of any material point (q), whose normal projection on mid-surface is the material point (p), in the initial states C_0 are given by

$$\boldsymbol{X}_q\left(\xi^1, \xi^2, z\right) = \boldsymbol{X}_p\left(\xi^1, \xi^2\right) + z\boldsymbol{D}\left(\xi^1, \xi^2\right), \quad z \in [-h/2, h/2] \tag{1}$$

where h is the thickness, $\boldsymbol{X}_p$ is a point of the reference surface, and $\boldsymbol{D}$ is the initial shell director. The base vectors, in the initial state C_0 are given as

$$\boldsymbol{G}_\alpha = \boldsymbol{A}_\alpha + z\boldsymbol{D}_\alpha; \quad \boldsymbol{G}_3 = \boldsymbol{D}, \quad \alpha = 1,2 \tag{2}$$

The element of surface, dA, in the initial state is given by

$$dA = \sqrt{A}\, dA_\xi, \qquad \sqrt{A} = \|\boldsymbol{A}_1 \wedge \boldsymbol{A}_2\|, \quad dA_\xi = d\xi^1 d\xi^2 \tag{3}$$

The covariant reference metric $\boldsymbol{G}$ at a material point $\boldsymbol{\xi}$ is defined by

$$\boldsymbol{G} = \left[\boldsymbol{G}_i \cdot \boldsymbol{G}_j\right], \quad i,j = 1,2,3 \tag{4}$$

With the hypothesis of a double directors shell model, the position vector, of any point q, in the deformed configuration is given by:

$$\boldsymbol{x}_q\left(\xi^1, \xi^2, z\right) = \boldsymbol{x}_p\left(\xi^1, \xi^2\right) + f_1(z)\boldsymbol{d}_1\left(\xi^1, \xi^2\right) + f_2(z)\boldsymbol{d}_2\left(\xi^1, \xi^2\right) \tag{5}$$

The base vectors, in the deformed state are then

$$\boldsymbol{g}_\alpha = \boldsymbol{a}_\alpha + f_1(z)\,\boldsymbol{d}_{1,\alpha} + f_2(z)\,\boldsymbol{d}_{2,\alpha}; \quad \boldsymbol{g}_3 = f_1'(z)\boldsymbol{d}_1 + f_2'(z)\boldsymbol{d}_2 \tag{6}$$

2.2 Strain Measure

The Lagrangian strain measure E, would have the following components:

$$\boldsymbol{E} = \frac{1}{2}(\boldsymbol{g} - \boldsymbol{G})$$
$$E_{ij} = \frac{1}{2}\left(g_{ij} - G_{ij}\right), \quad \begin{cases} E_{\alpha\beta} = e_{\alpha\beta} + f_1(z)\chi^1_{\alpha\beta} + f_2(z)\chi^2_{\alpha\beta} \\ 2E_{\alpha 3} = f_1'(z)\gamma^1_\alpha + f_2'(z)\gamma^2_\alpha \\ E_{33} = 1/2\left[\left(f_1' + f_2'\right)^2 d - 1\right] \end{cases} \quad g_{ij} = \boldsymbol{g}_i \cdot \boldsymbol{g}_j \tag{7}$$

where $e_{\alpha\beta}$ denote the membrane strains, $\chi^k_{\alpha\beta}$ the bending strains and γ^k_α the shear strains, which can be computed as

$$\begin{cases} e_{\alpha\beta} = \left(a_{\alpha\beta} - A_{\alpha\beta}\right)/2, \gamma^k_\alpha = c^k_\alpha - C^k_\alpha \\ \chi^k_{\alpha\beta} = \left(b^k_{\alpha\beta} - B^k_{\alpha\beta}\right)/2, k = 1,2 \end{cases} \tag{8}$$

where $a_{\alpha\beta}$, $b^k_{\alpha\beta}$ and c^k_α (k = 1, 2) represent, respectively, the covariant metric surface, the first curvature tensors and the shear:

$$\begin{cases} a_{\alpha\beta} = \boldsymbol{a}_\alpha \cdot \boldsymbol{a}_\beta, & c^k_\alpha = \boldsymbol{a}_\alpha \cdot \boldsymbol{d}_k \\ b^k_{\alpha\beta} = \boldsymbol{a}_\alpha \cdot \boldsymbol{d}_{k,\beta} + \boldsymbol{a}_\beta \cdot \boldsymbol{d}_{k,\alpha}, & d = \boldsymbol{d}_1 \cdot \boldsymbol{d}_1 \end{cases}, \quad k = 1,2 \tag{9}$$

The vector representation of membrane, bending and shear strains are given by:

$$\boldsymbol{e} = \begin{Bmatrix} e_{11} \\ e_{22} \\ 2e_{12} \end{Bmatrix}, \quad \chi^k = \begin{Bmatrix} \chi^k_{11} \\ \chi^k_{22} \\ 2\,\chi^k_{12} \end{Bmatrix}, \quad \gamma^k = \begin{Bmatrix} \gamma^k_1 \\ \gamma^k_2 \end{Bmatrix}, \quad k = 1,2 \tag{10}$$

To impose a third-order double director shell model, and at the same time a quadratic distribution of the shear stress, we chose the following expressions for $f_1(z)$ and $f_2(z)$:

$$f_1(z) = z - f_2(z), \quad f_2(z) = 4z^3/3h^2 \tag{11}$$

This gives the following shear strain:

$$2E_{\alpha 3} = f_1'(z)\gamma^1_\alpha + f_2'(z)\gamma^2_\alpha = \left(1 - 4\frac{z^2}{h^2}\right)\gamma^1_\alpha + 4\frac{z^2}{h^2}\gamma^2_\alpha \tag{12}$$

Vanishing of the transverse shear stress on the top and bottom shell faces, the shear strain can be obtained as follows: $\gamma^2_\alpha = 0$, $2E_{\alpha 3} = (1 - 4z^2/h^2)\gamma^1_\alpha$.

3 Weak Form and Linearization

The numerical solution with the finite element method is based on the weak form of equilibrium equations. The three dimensional form of the latter in the total Lagrangian formulation is given as

$$G = \int_V S^{ij}\delta E_{ij} dV - G_{ext} = 0 \tag{13}$$

where δE_{ij} are the covariant components of the virtual Green–Lagrange strain tensor, S^{ij} are the contravariant components of the second Piola–Kirchhoff stress tensor and G_{ext} is the external virtual work. Performing the integration through the thickness of the shell, and using Eq. (13), we get:

$$G = \int_A \left(\delta \boldsymbol{e} \cdot \boldsymbol{N} + \sum_{k=1}^{2} (\delta \boldsymbol{\chi}^k \cdot \boldsymbol{M}_k) + \delta\gamma^1 \cdot \boldsymbol{T}_1\right) dA - G_{ext} = 0 \tag{14}$$

where $\delta \boldsymbol{e}$, $\delta \boldsymbol{\chi}^k$ and $\delta \gamma^1$ are the variations of shell strain measures and $\boldsymbol{N}, \boldsymbol{M}_k$ and $\boldsymbol{T}_1$ are the membrane, bending and shear stresses resultants which can be written in matrix form as

$$\boldsymbol{N} = \begin{Bmatrix} N^{11} \\ N^{22} \\ N^{12} \end{Bmatrix}, \quad \boldsymbol{M}_k = \begin{Bmatrix} M_k^{11} \\ M_k^{22} \\ M_k^{12} \end{Bmatrix}, \quad \boldsymbol{T}_1 = \begin{Bmatrix} T_1^1 \\ T_1^2 \end{Bmatrix} \quad k = 1,2 \tag{15}$$

Their components are defined as follows

$$N^{\alpha\beta} = \int_{-h/2}^{h/2} S^{\alpha\beta} \sqrt{G/A} dz, \quad M_k^{\alpha\beta} = \int_{-h/2}^{h/2} f_k(z) S^{\alpha\beta} \sqrt{G/A} dz \tag{16}$$

$$T_1^{\alpha} = \int_{-h/2}^{h/2} f_1'(z) S^{\alpha 3} \sqrt{G/A}\, dz$$

As already indicated, the virtual strains can be obtained as the first variation of the strain measures which yields from Eq. (8):

$$\begin{cases} \delta e_{\alpha\beta} = 1/2 \left(\boldsymbol{a}_\alpha \cdot \delta \boldsymbol{x}_{,\beta} + \boldsymbol{a}_\beta \cdot \delta \boldsymbol{x}_{,\alpha} \right), \quad \delta \gamma_\alpha^1 = \boldsymbol{a}_\alpha \cdot \delta \boldsymbol{d}_k + \delta \boldsymbol{x}_{,\alpha} \cdot \boldsymbol{d}_k \\ \delta \chi_{\alpha\beta}^k = 1/2 \left(\boldsymbol{a}_\alpha \cdot \delta \boldsymbol{d}_{\mathrm{k},\beta} + \boldsymbol{a}_\beta \cdot \delta \boldsymbol{d}_{\mathrm{k},\alpha} + \delta \boldsymbol{x}_{,\alpha} \cdot \boldsymbol{d}_{\mathrm{k},\beta} + \delta \boldsymbol{x}_{,\beta} \cdot \boldsymbol{d}_{\mathrm{k},\alpha} \right) \end{cases} \tag{17}$$

Moreover, by defining the generalized resultant of stress and strain vectors with

$$\boldsymbol{R} = \begin{Bmatrix} \boldsymbol{N} \\ \boldsymbol{M}_1 \\ \boldsymbol{M}_2 \\ \boldsymbol{T}_1 \end{Bmatrix}_{11 \times 1}, \quad \boldsymbol{\Sigma} = \begin{Bmatrix} \boldsymbol{e} \\ \boldsymbol{\chi}^1 \\ \boldsymbol{\chi}^2 \\ \boldsymbol{\gamma}^1 \end{Bmatrix}_{11 \times 1} \tag{18}$$

the weak form of the equilibrium equation can be rewritten as

$$G(\boldsymbol{\Phi}, \delta \boldsymbol{\Phi}) = \int_A \delta \boldsymbol{\Sigma}^T \cdot \boldsymbol{R} dA - G_{ext}(\boldsymbol{\Phi}, \delta \boldsymbol{\Phi}) = 0 \tag{19}$$

where $\boldsymbol{\Phi} = (\boldsymbol{u}, \boldsymbol{d}_1, \boldsymbol{d}_2)$. Equation (19) defines the nonlinear shell problem, which can be solved by the Newton iterative procedure. The consistent tangent operator for the Newton solution procedure can be constructed by the directional derivative of the weak form in the direction of the increment $\Delta \boldsymbol{\Phi} = (\Delta \boldsymbol{u}, \Delta \boldsymbol{d}_1, \Delta \boldsymbol{d}_2)$. It is a conventional practice to split the tangent operator into geometric and material parts. The geometric part results from the variation of the virtual strains while holding stress resultants constant. This geometric part is not developed in this chapter. The material part of the tangent operator results from the variation in the stress resultants and thus takes the form:

$$D_M G \cdot \boldsymbol{\Delta\Phi} = \int_A \delta\boldsymbol{\Sigma}^T \cdot \boldsymbol{\Delta R} dA = \int_A \delta\boldsymbol{\Sigma}^T \cdot \boldsymbol{H}_T \cdot \Delta\boldsymbol{\Sigma} dA \tag{20}$$

where $\boldsymbol{H}_T$ is the material tangent modulus, is expressed as:

$$\boldsymbol{H}_T = \begin{bmatrix} \boldsymbol{H}_{11} & \boldsymbol{H}_{12} & \boldsymbol{H}_{13} & 0 \\ & \boldsymbol{H}_{22} & \boldsymbol{H}_{23} & 0 \\ & & \boldsymbol{H}_{33} & 0 \\ Sym & & & \boldsymbol{H}_c \end{bmatrix} \tag{21}$$

$$(\boldsymbol{H}_{11}, \boldsymbol{H}_{12}, \boldsymbol{H}_{13}, \boldsymbol{H}_{22}, \boldsymbol{H}_{23}, \boldsymbol{H}_{33}) = \int_{-h/2}^{h/2} (1, f_1, f_2, f_1^2, f_1 f_2, f_2^2) \boldsymbol{H} dz \tag{22}$$

$$\boldsymbol{H}_c = \int_{-h/2}^{h/2} (f_1')^2 \boldsymbol{H}_\tau dz \tag{23}$$

where $\boldsymbol{H}$ and $\boldsymbol{H}_\tau$ are in plane and out-of-plane linear elastic sub-matrices:

$$\boldsymbol{H} = \frac{E(z)}{1 - \nu^2(z)} \begin{bmatrix} 1 & \nu(z) & 0 \\ \nu(z) & 1 & 0 \\ 0 & 0 & (1 - \nu(z))/2 \end{bmatrix}, \quad \boldsymbol{H}_\tau = \frac{E(z)}{2(1 + \nu(z))} \begin{bmatrix} 1 & 0 \\ 0 & 1 \end{bmatrix} \tag{24}$$

In this chapter we consider an FGM shell structure made from mixture of the two constituents (for example: metal and ceramic), in which the composition is varied continuously in the thickness direction by the power-law distribution (P-FGM) expressed as:

$$P(z) = (P_c - P_m) V_c + P_m \quad \text{with } V_c(z) = \left(\frac{z}{h} + \frac{1}{2}\right)^n \tag{25}$$

where P denotes the effective material property, P_m and P_c represents the properties of the metal and ceramic, respectively, V_c is the volume fraction of the ceramic and n is the power-law index.

4 Finite Element Approximation

In this section, we elaborate the numerical implementation of the presented shell theoretical formulation based upon a four node shell element. The displacement vector is defined as: $\boldsymbol{u} = \boldsymbol{x}_p - \boldsymbol{X}_p$. Using the isoparametric concept, the variation and incremental of displacement vector is approximated by

$$\delta \boldsymbol{u} = \sum_{I=1}^{4} N^I \delta \boldsymbol{u}_I, \quad \Delta \boldsymbol{u} = \sum_{I=1}^{4} N^I \Delta \boldsymbol{u}_I \tag{26}$$

where N^I are the standard isoparametric shape functions. For further details concerning isoparametic concept, we refer to standard references (Dhatt and Touzot 1981) among authors. The first director vector $\boldsymbol{d}_1$ is approximated with the same functions:

$$\delta \boldsymbol{d}_1 = \sum_{I=1}^{4} N^I \delta \boldsymbol{d}_{1I}, \quad \Delta \boldsymbol{d}_1 = \sum_{I=1}^{4} N^I \Delta \boldsymbol{d}_{1I} \tag{27}$$

4.1 Membrane and First Bending Strain Field

We first consider the shell membrane part of the problem. The strain-displacement relation is:

$$\delta \boldsymbol{e} = \boldsymbol{B}_m \cdot \delta \boldsymbol{\Phi}_n \tag{28}$$

where $\boldsymbol{B}_m$ is the discrete displacement approximation to the membrane strain displacement operator defined as

$$\boldsymbol{B}_m^I = \begin{bmatrix} \boldsymbol{B}_{mm}^I & \boldsymbol{0} & \boldsymbol{0} \end{bmatrix}, \quad \boldsymbol{B}_{mm}^I = \begin{bmatrix} \boldsymbol{a}_1^T N_1^I \\ \boldsymbol{a}_2^T N_2^I \\ \boldsymbol{a}_1^T N_2^I + \boldsymbol{a}_2^T N_1^I \end{bmatrix} \tag{29}$$

For the first bending part, the strain-displacement relation is given by

$$\delta \chi^1 = \boldsymbol{B}_1 \cdot \delta \boldsymbol{\Phi}_n \tag{30}$$

where $\boldsymbol{B}_1$ is the discrete first bending strain-displacement operator:

$$\boldsymbol{B}_1^I = \begin{bmatrix} \boldsymbol{B}_{1m}^I & \boldsymbol{B}_{1b}^I & \boldsymbol{0} \end{bmatrix}, \quad \boldsymbol{B}_{1m}^I = \begin{bmatrix} \boldsymbol{d}_{1,1}^T N_1^I \\ \boldsymbol{d}_{1,2}^T N_2^I \\ \boldsymbol{d}_{1,1}^T N_2^I + \boldsymbol{d}_{1,2}^T N_1^I \end{bmatrix}, \quad \boldsymbol{B}_{1b}^I = \begin{bmatrix} \boldsymbol{a}_1^T N_1^I \\ \boldsymbol{a}_2^T N_2^I \\ \boldsymbol{a}_1^T N_2^I + \boldsymbol{a}_2^T N_1^I \end{bmatrix} \tag{31}$$

4.2 Construction of the Assumed Natural Transverse Shear Strain Field

In order to eliminate the shear locking effect, the transverse shear strains are interpolated over a parent element by using the assumed natural strain (ANS) concept of Bathe and Dvorkin (1985) according to

$$\delta\gamma^1 = \begin{bmatrix} \delta\gamma_1^1 \\ \delta\gamma_2^1 \end{bmatrix} = \begin{bmatrix} (1-\eta)\,\delta\gamma_1^1(B) + (1-\eta)\,\delta\gamma_1^1(D) \\ (1-\xi)\,\delta\gamma_2^1(A) + (1-\xi)\,\delta\gamma_2^1(C) \end{bmatrix} \tag{32}$$

where A, B, C and D are the shear interpolation points at the midpoints of the element boundaries: (4, 1) (1, 2) (2, 3) and (3, 4) respectively. Finally, the strain-displacement relation is then given by:

$$\delta\gamma^1 = \boldsymbol{B}_s \cdot \delta\boldsymbol{\Phi}_n \tag{33}$$

where $\boldsymbol{B}_s$ is the discrete shear strain-displacement operator:

$$\boldsymbol{B}_s = \begin{bmatrix} N_1^1\boldsymbol{d}_{1B}^T & N_1^2\boldsymbol{a}_{1B}^T & \boldsymbol{0} & N_1^2\boldsymbol{d}_{1B}^T & N_1^2\boldsymbol{a}_{1B}^T & \boldsymbol{0} & N_1^3\boldsymbol{d}_{1D}^T & N_1^3\boldsymbol{a}_{1D}^T & \boldsymbol{0} & N_1^4\boldsymbol{d}_{1D}^T & N_1^3\boldsymbol{a}_{1D}^T & \boldsymbol{0} \\ N_2^1\boldsymbol{d}_{1A}^T & N_2^4\boldsymbol{a}_{2A}^T & \boldsymbol{0} & N_2^2\boldsymbol{d}_{1C}^T & N_2^3\boldsymbol{a}_{2C}^T & \boldsymbol{0} & N_2^3\boldsymbol{d}_{1C}^T & N_2^3\boldsymbol{a}_{2C}^T & \boldsymbol{0} & N_2^4\boldsymbol{d}_{1A}^T & N_2^4\boldsymbol{a}_{2A}^T & \boldsymbol{0} \end{bmatrix} \tag{34}$$

4.3 Discrete Constraints

The shear part relative to the second director vector $\boldsymbol{d}_2$ will be vanished in a discrete form, we choose a quadratic interpolation as the same one proposed in Dammak et al. (2005), to formulate a nonlinear discrete Kirchhoff four nodes shell element:

$$\delta\boldsymbol{d}_2 = \sum_{I=1}^{4} N^I \delta\boldsymbol{d}_{2I} + \sum_{K=5}^{8} P_K \delta\alpha_K \boldsymbol{t}_K\,,\quad \Delta\boldsymbol{d}_2 = \sum_{I=1}^{4} N^I \Delta\boldsymbol{d}_{2I} + \sum_{K=5}^{8} P_K \Delta\alpha_K \boldsymbol{t}_K \tag{35}$$

where (*I*) represent a node of the element (*K*) represent the mid-point of the element boundaries and $\delta\alpha_K$ are variables associated to $\delta\boldsymbol{d}_2$ at point (*K*) and $\boldsymbol{t}_K$ is a unit vector and its direction is defined by the position of the nodes couple (*I*, *J*).

$$\boldsymbol{t}_K = (\boldsymbol{x}_J - \boldsymbol{x}_I)/L_K, \quad L_K = \|\boldsymbol{x}_J - \boldsymbol{x}_I\| \tag{36}$$

When one introduces the vanishing shearing hypothesis, on top and bottom faces, over the element boundaries under integral form, we have for side (*I*, *J*):

$$\int_I^J \delta\gamma_{sz}^2 ds = 0, \tag{37}$$

After all made calculus (Dammak et al. 2005), the second bending strain is expressed as

$$\delta\boldsymbol{\chi}^2 = \boldsymbol{B}_2 \cdot \delta\boldsymbol{\Phi}_n \tag{38}$$

where $\boldsymbol{B}_2$ is the discrete second bending strain-displacement operator:

$$\boldsymbol{B}_2^I = \begin{bmatrix} \boldsymbol{B}_{2m}^I & \boldsymbol{0} & \boldsymbol{B}_{2b}^I \end{bmatrix} \tag{39}$$

$$\boldsymbol{B}_{2m}^I = \begin{bmatrix} \boldsymbol{d}_{2,1}^t N_1^I + \boldsymbol{a}_1^t \cdot \boldsymbol{M}_{d,1}^I \\ \boldsymbol{d}_{2,2}^t N_2^I + \boldsymbol{a}_2^t \cdot \boldsymbol{M}_{d,2}^I \\ \boldsymbol{d}_{2,1}^t N_2^I + \boldsymbol{d}_{2,2}^t N_1^I + \boldsymbol{a}_1^t \cdot \boldsymbol{M}_{d,2}^I + \boldsymbol{a}_2^t \cdot \boldsymbol{M}_{d,1}^I \end{bmatrix}, \quad \boldsymbol{B}_{2b}^I = \begin{bmatrix} \boldsymbol{a}_1^t \cdot \boldsymbol{M}_{r,1}^I \\ \boldsymbol{a}_2^t \cdot \boldsymbol{M}_{r,2}^I \\ \boldsymbol{a}_1^t \cdot \boldsymbol{M}_{r,2}^I + \boldsymbol{a}_2^t \cdot \boldsymbol{M}_{r,1}^I \end{bmatrix} \tag{40}$$

Finally, the generalized strain $\delta\boldsymbol{\Sigma}$ can be expressed as follows:

$$\delta\boldsymbol{\Sigma} = \begin{Bmatrix} \delta\boldsymbol{e} \\ \delta\boldsymbol{\chi}^1 \\ \delta\boldsymbol{\chi}^2 \\ \delta\boldsymbol{\gamma}^1 \end{Bmatrix} = \boldsymbol{B} \cdot \delta\boldsymbol{\Phi}_n, \quad \boldsymbol{B} = \begin{bmatrix} \boldsymbol{B}_m \\ \boldsymbol{B}_1 \\ \boldsymbol{B}_2 \\ \boldsymbol{B}_s \end{bmatrix} \tag{41}$$

4.4 Nodal Transformation

In all equations, $\delta\boldsymbol{d}_k$ and $\delta\boldsymbol{d}_{k,\alpha}$ are the variation of the directors and there derivatives. These variations can be written either in spatial description

$$\delta\boldsymbol{d}_k = \delta\boldsymbol{\theta}_k \wedge \boldsymbol{d}_k = \boldsymbol{\Lambda}_k\, \delta\boldsymbol{\theta}_k, \quad \overline{\Lambda}_k = -\tilde{\boldsymbol{d}}_k \tag{42}$$

where $\tilde{\boldsymbol{d}}_k$ is the skew-symmetric tensor such that $\tilde{\boldsymbol{d}}_k \boldsymbol{d}_k = \boldsymbol{0}$, or in material description

$$\delta \boldsymbol{d}_k = \boldsymbol{Q}_k \, \delta \bar{\boldsymbol{\Theta}}_k \, \boldsymbol{E}_3 = \bar{\boldsymbol{\Lambda}}_k \delta \boldsymbol{\Theta}_k, \quad \bar{\boldsymbol{\Lambda}}_k = \boldsymbol{Q}_k \tilde{\boldsymbol{E}}_3 \tag{43}$$

where we assumed that $\boldsymbol{d}_k = \boldsymbol{Q}_k \boldsymbol{E}_3$ and $\boldsymbol{E}_3 = \begin{bmatrix} 0 & 0 & 1 \end{bmatrix}^t$. A spatial description leads to a shell problem with nine DOF/node and the material description leads to a shell problem with 7 DOF/node, where the transformation $\bar{\boldsymbol{\Lambda}}$ take the following form:

$$\bar{\boldsymbol{\Lambda}}_k = \begin{bmatrix} -\boldsymbol{t}_{2k} & \boldsymbol{t}_{1k} \end{bmatrix}_{3\times 2} \tag{44}$$

5 Enhanced Membrane Strain

To improve the membrane behavior of the bilinear shell element, especially for in-plane bending dominated case; we enhance the compatible in-plane strain with a four parameters field $\boldsymbol{\alpha} = \begin{bmatrix} \alpha_1 & \alpha_2 & \alpha_3 & \alpha_4 \end{bmatrix}^T$:

$$\boldsymbol{e} = \boldsymbol{e}^c + \boldsymbol{e}^{inc} \tag{45}$$

$$\delta e^c = \boldsymbol{B}_m \cdot \delta \boldsymbol{\Phi}_n, \quad e^{inc} = \tilde{\boldsymbol{B}}_m \cdot \boldsymbol{\alpha}, \quad \delta e^{inc} = \tilde{\boldsymbol{B}}_m \cdot \delta \boldsymbol{\alpha} \tag{46}$$

The orthogonality condition is expressed as:

$$\int\limits_A \boldsymbol{e}^{inc\,T} \cdot \boldsymbol{N} dA = 0 \tag{47}$$

With this enhancement and the orthogonality condition, the variation of the three fields functional is written as:

$$G = \int\limits_A \left(\delta \boldsymbol{e}^T \cdot \boldsymbol{N} + \sum_{k=1}^{2} \left(\delta \boldsymbol{\chi}^{kT} \cdot \boldsymbol{M}_k \right) + \delta \gamma^{1T} \cdot \boldsymbol{T}_1 \right) dA - G_{ext} = 0 \tag{48}$$

Further, after local condensation of parameter $\boldsymbol{\alpha}$, we obtain:

$$\boldsymbol{K}_d = \int_A \boldsymbol{B}^T \mathbf{H}_T \boldsymbol{B} dA \tag{49}$$

$$\boldsymbol{b} = \int\limits_A \tilde{\boldsymbol{B}}_m^T (\boldsymbol{H}_{11} \boldsymbol{B}_m + \boldsymbol{H}_{12} \boldsymbol{B}_1 + \boldsymbol{H}_{13} \boldsymbol{B}_2) dA, \quad \boldsymbol{a} = \int\limits_A \tilde{\boldsymbol{B}}_m^T \boldsymbol{H}_{11} \tilde{\boldsymbol{B}}_m dA \tag{50}$$

Table 1 Listing of shell elements

Name	Description
MIXED	FSDT, Simo et al. (1989) with Hellinger–Reissner formulation
SHO4	Present double directors shell element: displacement formulation
SHO4I4	Present double directors shell element: enhanced formulation

$$\Delta\boldsymbol{\alpha} = -\boldsymbol{a}^{-1}(\boldsymbol{R}_{\alpha} + \boldsymbol{b}\cdot\Delta\boldsymbol{\Phi}_n), \quad \boldsymbol{R}_{\alpha} = \int_A \tilde{\boldsymbol{B}}_m^T \cdot \boldsymbol{N}dA \tag{51}$$

The contribution of the element material tangent stiffness can then be computed as

$$\boldsymbol{K}_M = \boldsymbol{K}_d - \boldsymbol{b}^T\boldsymbol{a}^{-1}\boldsymbol{b} \tag{52}$$

And the residual:

$$\boldsymbol{R} = \boldsymbol{f}_u - \left(\boldsymbol{R}_u - \boldsymbol{b}^T\boldsymbol{a}^{-1}\boldsymbol{R}_{\alpha}\right), \quad \boldsymbol{R}_u = \int_A \boldsymbol{B}^T \cdot \boldsymbol{R}dA \tag{53}$$

Finally, the enhancement matrix $\tilde{\boldsymbol{B}}_m$, is given by :

$$\tilde{\boldsymbol{B}}_m = \frac{\sqrt{A_0}}{\sqrt{A}}\boldsymbol{T}_0^{-T}\boldsymbol{M}, \quad \boldsymbol{M} = \begin{bmatrix} \xi & 0 & 0 & 0 \\ 0 & \eta & 0 & 0 \\ 0 & 0 & \xi & \eta \end{bmatrix} \tag{54}$$

6 Numerical Examples

The performance of the proposed discrete double director shell element with enhanced membrane strain is evaluated with several problems. The convergence of the results is compared to other well-known formulations. A listing of these shell elements, and the abbreviations used to identify them henceforth, is contained in Table 1.

6.1 Cook's Membrane Problem

A trapezoïdal plate is clamped on one end and subjected to a distributed in-plane bending load on the other end, as shown in Fig. 1. This problem has a considerable amount of shear deformation, and is an excellent test of an element to model

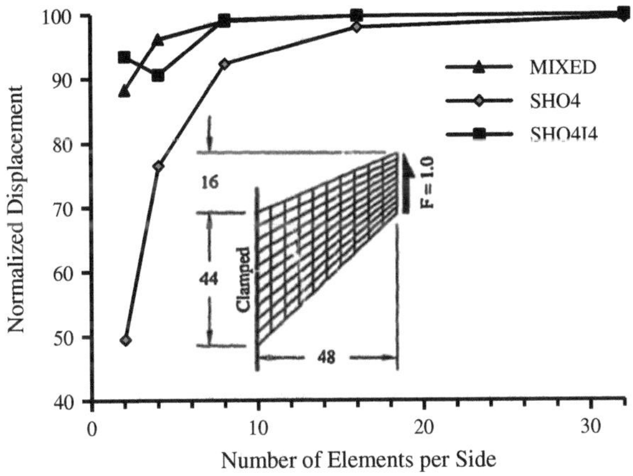

Fig. 1 Description and results of Cook's membrane problem

Table 2 Results of Cook's membrane problem

Mesh	MIXED		SHO4		SHO4I4	
	Val.	(%)	Val.	(%)	Val.	(%)
2 × 2	21.124	88.2	11.84	49.4	22.37	93.4
4 × 4	23.018	96.1	18.30	76.4	21.66	90.5
8 × 8	23.685	98.9	22.08	92.2	23.72	99.1
16 × 16	23.878	99.7	23.43	97.9	23.88	99.7
32 × 32	–	–	23.81	99.4	23.93	99.9

Table 3 Center deflections of isotropic homogeneous plates

h	Classic, Timoshenko	Zenkour (2006)	SHO4, SHO4I4
0.01	44360.9	44383.84	44375
0.03	1643.00	1650.646	1650
0.1	44.3609	46.6581	46.65

membrane dominated situations with skewed meshes. The material properties are: E $= 1.0$, $\nu = 0.33$, and $h = 1.0$. A finite element converged solution of 23.94, obtained with a 16 nodes element with 24 × 24 meshing, is used to normalize the results, which are shown in Fig. 1 and listed in Table 2.

Table 4 Distribution of stresses across the thickness of isotropic homogeneous plates

h	Coordinate	$\sigma_{xx}(0.5a, 0.5b, z)$ Zenkour	$\sigma_{xx}(0.5a, 0.5b, z)$ SHO4, SHO4I4	$\sigma_{xy}(0, 0, -z)$ Zenkour	$\sigma_{xy}(0, 0, -z)$ SHO4, SHO4I4
0.01	0.005	2873.39	2880	1949.36	1926
	0.004	2298.57	2302	1559.04	1541
	0.003	1723.84	1725	1168.99	1156
	0.002	1149.18	1150	779.18	771.1
	0.001	574.58	574.6	389.55	385.6
	0.000	0.000	0.000	0.000	0.000
0.03	0.005	319.445	320.2	217.156	214.4
	0.004	255.415	255.8	173.282	171.3
	0.003	191.472	191.6	129.682	128.4
	0.002	127.603	127.7	86.313	85.52
	0.001	63.788	63.80	43.112	42.74
	0.000	0.000	0.000	0.000	0.000
0.1	0.005	28.9307	28.99	20.0476	19.67
	0.004	23.0055	23.03	15.6459	15.45
	0.003	17.1660	17.19	11.4859	11.42
	0.002	11.3994	11.41	7.5315	7.537
	0.001	5.6858	5.694	3.7265	3.745
	0.000	0.000	0.000	0.000	0.000

6.2 Simply Supported Isotropic Homogeneous Plates Under Uniformly Distributed Load

In order to prove the validity of the present formulation, results were obtained for isotropic square plates ($a = b = 1$) under uniformly distributed load ($q = 1$). The material properties are: $(E_c = E_m = 1,\ \nu = 0.3)$ and three values for the plate thickness: $h = 0.01, 0.03$, and 0.1. The numerical results are presented in Tables 3 and 4 Mesh used is 20×20.

It is to be noted that the present results compare very well and show good convergence with the exact solution of Zenkour (2006).

6.3 Simply Supported FGM Square Plates Under Uniformly Distributed Load

Now, a functionally graded material consisting of aluminum and alumina is considered. The material properties are: $(E_c = 380 \times 10^9,\ E_m = 70 \times 10^9,\ \nu = 0.3)$, $a = b = 1$ and plate thickness: $h = 0.1$. Mesh used is 20×20. The various non-dimensional parameters used are:

Table 5 Effects of volume fraction exponent on the dimensionless stresses and displacements of a FGM square plate: uniform distributed loads ($a/h = 10$)

	$\bar{w}$_Z	$\bar{w}$_P	$\bar{\sigma}_x$_Z	$\bar{\sigma}_x$_P	$\bar{\sigma}_y$_Z	$\bar{\sigma}_y$_P	$\bar{\tau}_{xy}$_Z	$\bar{\tau}_{xy}$_P	$\bar{\tau}_{xz}$_Z	$\bar{\tau}_{xz}$_P	$\bar{\tau}_{yz}$_Z	$\bar{\tau}_{yz}$_P
Ce.	0.4665	0.4665	2.8932	2.899	1.9103	1.913	1.2850	1.275	0.5114	0.4697	0.4429	0.4175
1	0.9287	0.9286	4.4745	4.483	2.1692	2.171	1.1143	1.104	0.5114	0.4698	0.5446	0.5134
2	1.1940	1.1937	5.2296	5.238	2.0338	2.035	0.9907	0.9819	0.4700	0.4302	0.5734	0.5387
3	1.3200	1.3194	5.6108	5.619	1.8593	1.86	1.0047	0.9963	0.4367	0.3981	0.5629	0.5267
4	1.3890	1.3881	5.8915	5.9	1.7197	1.721	1.0298	1.021	0.4204	0.3823	0.5346	0.499
5	1.4356	1.4346	6.1504	6.159	1.6104	1.612	1.0451	1.037	0.4177	0.3795	0.5031	0.4691
6	1.4727	1.4716	6.4043	6.413	1.5214	1.523	1.0536	1.045	0.4227	0.3841	0.4755	0.4435
7	1.5049	1.5039	6.6547	6.663	1.4467	1.448	1.0589	1.051	0.4310	0.392	0.4543	0.424
8	1.5343	1.5333	6.8999	6.908	1.3829	1.384	1.0628	1.055	0.4399	0.4005	0.4392	0.4104
9	1.5617	1.5608	7.1383	7.147	1.3283	1.329	1.0662	1.058	0.4481	0.4085	0.4291	0.4015
10	1.5876	1.5867	7.3689	7.377	1.2820	1.283	1.0694	1.061	0.4552	0.4155	0.4227	0.396
Me.	2.5327	2.5327	2.8932	2.899	1.9103	1.913	1.2850	1.275	0.5114	0.4697	0.4429	0.4175

$$\bar{w} = \frac{10\,h^3 E_c}{q_0\,a^4} w_0\left(\frac{a}{2}, \frac{b}{2}\right), \bar{\sigma}_x = \frac{h}{a\,q_0}\sigma_x\left(\frac{a}{2}, \frac{b}{2}, \frac{h}{2}\right), \bar{\sigma}_y = \frac{h}{a\,q_0}\sigma_y\left(\frac{a}{2}, \frac{b}{2}, \frac{h}{3}\right),$$
$$\bar{\tau}_{xy} = \frac{h}{a\,q_0}\tau_{xy}\left(0, 0, -\frac{h}{3}\right), \bar{\tau}_{yz} = \frac{h}{a\,q_0}\tau_{yz}\left(\frac{a}{2}, 0, \frac{h}{6}\right), \bar{\tau}_{xz} = \frac{h}{a\,q_0}\tau_{xz}\left(0, \frac{b}{2}, 0\right)$$

In Table 5, the effect of volume fraction exponent on the dimensionless stresses and displacements of a FGM square plate (a/h = 10) is given. This table shows comparison between results for plates subjected to uniform distributed loads: (we denote by _Z the results of Zenkour (2006) and _P the present SHO4 and SHO4I4 formulations).

7 Conclusions

An enhanced double-directors model for FGM shells which includes the introduction of kinematic constrain has been presented. The kinematic constrain is imposed in a discrete form in the finite element approximation. The compatible in-plane strain is enhanced by using four parameters. We can conclude that the use of enhanced DDDSM model improve the membrane behavior of the bilinear shell element. The enhanced effects can be clearly observed when studying the static behavior of 3d-shell structures e.g. cylinder and sphere. Dimensionless stresses and displacements of the simply supported functionally graded plate under uniform and sinusoidal loading are computed by the present enhanced DDDSM model. Finally, we have shown by computed results that the stress and displacement components are predicted with high accuracy compared to available published exact results.

References

Başar Y, Itskov M, Eckstein A (2000) Composite laminates: nonlinear interlaminar stress analysis by multi-layer shell elements. Comput Methods Appl Mech Eng 185:367–397

Brank B (2005) Non linear shell models with seven kinematic parameters. Comput Methods Appl Mech Eng 194:2336–2362

Brank B, Carrera E (2000) A family of shear-deformable shell finite elements for composite structures. Comput Struct 76:287–297

Brank B, Korelc J, Ibrahimbegovic A (2002) Nonlinear shell problem formulation accounting for through-the-thickness stretching and its finite element implementation. Comput Struct 80:699–717

Bathe KJ, Dvorkin E (1985) A four-node plate bending element based on Mindlin/Reissner plate theory and a mixed interpolation. Int J Numer Meth Eng 21:367–383

Carrera E, Brischetto S, Cinefra M, Soave M (2011) Effects of thickness stretching in functionally graded plates and shells. Compos B 42:123–133

Castellazzi G, Gentilini C, Krysl P, Elishakoff I (2013) Static analysis of functionally graded plates using a nodal integrated finite element approach. Compos Struct 103:197–200

Dammak F, Abid S, Gakwaya A, Dhatt G (2005) A formulation of the non linear discrete Kirchhoff quadrilateral shell element with finite rotations and enhanced strains. Eur J Comput Mech 14:1–26

Dhatt G, Touzot G (1981) Une présentation de la méthode des éléments finis, Maloine SA (ed). Paris et Les Presses de l'Université Laval, Québec

Ferreira AJM, Roque CMC, Jorge RMN, Fasshaueret GE, Batra RC (2007) Analysis of functionally graded plates by a robust meshless method. Mech Adv Mater Struct 14(8): 577–587

Nguyen TT, Sab K, Bonnet G (2008) First-order shear deformation plate models for functionally graded materials. Compos Struct 83:25–36

Reddy JN (2000) Analysis of functionally graded plates. Int J Numer Meth Eng 47:663–684

Simo JC, Fox DD, Rifai MS (1989) On a stress resultant geometrically exact shell model. Part II: the linear theory; computational aspects. Comp Methods Appl Mech Eng 73:53–92

Simo JC, Rifai MS (1990) A class of mixed assumed strain methods and the method of incompatible modes. Int J Numer Meth Eng 29:1595–1638

Singha MK, Prakash T, Ganapathi M (2011) Finite element analysis of functionally graded plates under transverse load. Finite Elem Anal Des 47:453–460

Xiang S, Kang GW (2013) A nth-order shear deformation theory for the bending analysis on the functionally graded plates. Eur J Mech A/Solids 37:336–343

Xiao-Hong Wu, Chen Changqing, Shen Ya-Peng, Tian Xiao-Geng (2002) A high order theory for functionally graded piezoelectric shells. Int J Solids Struct 39:5325–5344

Zenkour AM (2006) Generalized shear deformation theory for bending analysis of functionally graded plates. Appl Math Model 30:67–84

Modal Analysis of Helical Planetary Gear Train Coupled to Bevel Gear

Maha Karray, Nabih Feki, Fakher Chaari and Mohamed Haddar

Abstract The purpose of this work is to determine the critical frequencies of planetary gear train connected to bevel gear. A three dimensional model of helical planetary gear is proposed where 6 degrees of freedom per component are considered while 4 degrees of freedom (3 translations and 1 rotation) per component of bevel gear is adopted. Connection between the two stages is done by a torsional and linear springs. The governing equation of motion is derived. The mean values of both mesh stiffness are used to solve the eigenvalue problem. The critical natural frequencies for the system are determined. It was found that the natural modes can be classified into two major classes depending on their modal displacements.

Keywords Planetary gear · Bevel gear · Three dimensional model · Natural frequencies · Natural modes

1 Introduction

Gear transmissions are widely used in a wide number of machines and vehicles taking advantage from their advantages such as power-to-weight ratio, reduced cost and high efficiency. This chapter investigates modal characteristics of gearbox

M. Karray (✉) · N. Feki · F. Chaari · M. Haddar
Laboratory of Mechanics Modelling and Prod.ustion, National School of Engineers of Sfax, BP 1173, 3038 Sfax, Tunisia
e-mail: maha-karray@hotmail.fr

F. Chaari
e-mail: fakher.chaari@gmail.com

M. Haddar
e-mail: mohamed.haddar@enis.rnu.tn

M. S. Abbes et al. (eds.), *Mechatronic Systems: Theory and Applications*,
Lecture Notes in Mechanical Engineering, DOI: 10.1007/978-3-319-07170-1_14,

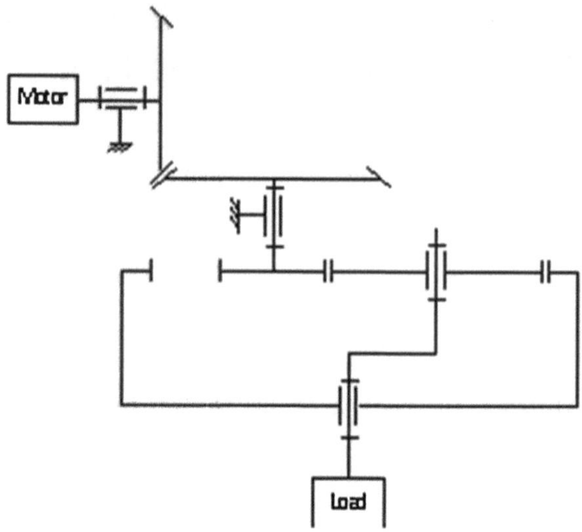

Fig. 1 Model of bucket wheel excavator

composed of two stages. The first one is a spiral bevel gear and the second is a helical planetary gearbox as shown in Fig. 1.

Planetary gear sets consist of either spur or helical gears. Spur planetary gear sets can be commonly found in heavy machinery and off-highway gearboxes and transmissions, while the helical planetary gear sets are the norm for all automotive applications as in automatic transmissions and transfer cases.

Bevel gears transmissions are widely used in automotive differentials and aerospace applications for their ability to transmit torque between non-parallel shafts. The most common of these are straight bevel gears and spiral bevel gears.

The modelling of the vibratory behaviour of planetary gear was widely treated in literature (Hidaka et al. 1980). Most of the models employed two dimensional (2D) formulations, which can only consider spur gears. Lin and Parker (1999) recovered for this kind of models three types of modes: translational, rotational and planet modes.

However, helical gears, which are shown to be quite different dynamically from spur gears (Kahraman 1994a, b), are generally preferred since they are quieter especially in automotive applications.

The modelling of the vibratory behaviour of parallel axis geared rotor system was widely treated in literature (Ozguven and Houser 1988a, b; Maatar and Velex 1995) however few research works were dedicated to bevel gears dynamics. Li et al. (2010) present an 8 DOF degrees of freedom nonlinear dynamic model of a spiral bevel gear pair which involves time-varying mesh stiffness, transmission error, backlash, and asymmetric mesh stiffness. Li et al. (1998) proposed a new method to perform the static analysis of straight and helical bevel gears by finite element methods.

Bartelmus and Zimroz (2011) classified gearbox into compound and complex gearboxes where we can find both bevel and planetary gears and they determined their characteristic frequencies. Bartelmus (2011) used the same classifications to make a diagnostic feature.

This chapter will investigate a dynamic model of a bevel gear transmission coupled to a single stage helical planetary gearbox. The objective is to characterize the modal properties of such transmission.

2 Model and Equation of Motion

2.1 *Helical Planetary Gear*

We are treating a three-dimensional vibration problem of a one-stage helical planetary gear train consisting of a sun gear (s), a ring gear (r), which are coupled to each other by n planets (P_i) mounted on a carrier (c) and are considered as rigid bodies. Bearings are modelled by linear springs. Each mesh stiffness is modelled by linear spring acting on the line of action (Hbaieb et al. 2005).

First, the equation of motion of each component is derived separately, and then assembled to obtain the overall system matrices of an n-planet helical planetary gear train. Each component has six degrees of freedom: three translations (u_j, v_j and w_j) and three rotations (φ_j, ψ_j and θ_j, $j = c, r, s, 1 \ldots n$). These coordinates are measured with respect to a frame $(O, \overrightarrow{s_1}, \overrightarrow{t_1}, \overrightarrow{z_1})$ fixed to the carrier and rotating with a constant angular speed Ω_c.

The rotations (φ_j, ψ_j and θ_j) are replaced by their corresponding translational gear mesh displacements as:

$$\rho_{jx} = Rb_j\varphi_j,\ \rho_{jy} = Rb_j\psi_j,\ \rho_{jz} = Rb_j\theta_j, \quad j = c, r, s, 1, .., n \tag{1}$$

where Rb_j is the base circle radius for the sun, ring, planet, and the radius of the circle passing through planet centers for the carrier. Circumferential planet locations are specified by the fixed angles α_i, which is measured relative to the rotating basis vector $\overrightarrow{s_1}$ so that $\alpha_1 = 0$.

The displacement vector q_j is defined as:

$$q_{jPG} = \left[u_j,\ v_j,\ w_j,\ \rho_{jx},\ \rho_{jx},\ \rho_{jx}\right]^T \tag{2}$$

The relative radial Δ_{ir}, tangential Δ_{it} and axial Δ_{iz}, displacements of planet i on the bearing are defined as (Hbaieb et al. 2005):

Table 1 The coefficients of Eqs. (6) and (7)

$s_{1i} = \cos\beta\sin(\alpha_S - \alpha_i)$	$r_{1i} = -\cos\beta\sin(\alpha_r + \alpha_i)$
$s_{2i} = \cos\beta\cos(\alpha_S - \alpha_i)$	$r_{2i} = \cos\beta\cos(\alpha_r + \alpha_i)$
$s_{3i} = \sin\beta$	$r_{3i} = -\sin\beta$
$s_{4i} = -\sin\beta\sin(\alpha_S - \alpha_i) + \frac{1}{Rb_S} p_1 \sin\beta\cos(\alpha_S - \alpha_i)$	$r_{4i} = -\sin\beta\sin(\alpha_r + \alpha_i) + \frac{1}{Rb_C} p'_2 \sin\beta\cos(\alpha_r + \alpha_i)$
$s_{5i} = -\sin\beta\cos(\alpha_S - \alpha_i) - \frac{1}{Rb_S} p_1 \sin\beta\sin(\alpha_S - \alpha_i)$	$r_{5i} = \sin\beta\cos(\alpha_r + \alpha_i) - \frac{1}{Rb_C} p'_2 \sin\beta\sin(\alpha_r + \alpha_i)$
$s_{6i} = \cos\beta$	$r_{6i} = \cos\beta$
$s_{7i} = -\cos\beta\sin\alpha_S$	$r_{7i} = \cos\beta\sin(\alpha_r)$
$s_{8i} = -\cos\beta\cos\alpha_S$	$r_{8i} = -\cos\beta\cos(\alpha_r)$
$s_{9i} = -\sin\beta$	$r_{9i} = \sin\beta$
$s_{10i} = -\sin\beta\sin\alpha_S + \frac{1}{Rb_{St}} p_2 \sin\beta\cos\alpha_S$	$r_{10i} = \sin\beta\sin(\alpha_r) + \frac{1}{Rb_{st}} P'_1 \sin\beta\cos(\alpha_r)$
$s_{11i} = -\sin\beta\cos\alpha_S - \frac{1}{Rb_{St}} p_2 \sin\beta\sin\alpha_S$	$r_{11i} = -\sin\beta\cos(\alpha_r) + \frac{1}{Rb_{st}} P'_1 \sin\beta\sin(\alpha_r)$
$s_{12i} = \cos\beta$	$r_{12i} = -\cos\beta$
$p_1 = Rb_s tg(\alpha_s)$	$p'_1 = Rb_r tg(\alpha_r)$
$p_2 = Rb_i tg(\alpha_s)$	$p'_2 = Rb_i tg(\alpha_r)$

$$\Delta_{ir} = u_c\cos\alpha_i + v_c\sin\alpha_i - u_i \tag{3}$$

$$\Delta_{it} = -u_c\sin\alpha_i + v_c\cos\alpha_i + \rho_{cz} - v_i \tag{4}$$

$$\Delta_{iz} = w_c + \rho_{cx}\sin\alpha_i - \rho_{cy}\cos\alpha_i - w_i \tag{5}$$

The relative gear mesh displacements at the line of contact for the sun gear meshing with planet i Δ_{si} and for the planet i meshing with the ring Δ_{ri} are defined as follow (Saada et al. 1992):

$$\begin{aligned}\Delta_{si} = {} & s_{1i}u_s + s_{2i}v_s + s_{3i}w_s + s_{4i}\rho_{sx} + s_{5i}\rho_{sy} + s_{6i}\rho_{sz} + s_{7i}u_i \\ & + s_{8i}v_i + s_{9i}w_i + s_{10i}\rho_{ix} + s_{11i}\rho_{iy} + s_{12i}\rho_{iz}\end{aligned} \tag{6}$$

$$\begin{aligned}\Delta_{ri} = {} & r_{1i}u_r + r_{2i}v_r + r_{3i}w_r + r_{4i}\rho_{rx} + r_{5i}\rho_{ry} + r_{6i}\rho_{rz} + r_{7i}u_i \\ & + r_{8i}v_i + r_{9i}w_i + r_{10i}\rho_{ix} + r_{11i}\rho_{iy} + r_{12i}\rho_{iz}\end{aligned} \tag{7}$$

The coefficients of Eqs. (6) and (7) are given in Table 1.

Where β is the helix angle, α_r and α_s are the pressure angles of the mesh ring-planet and sun-planet.

Applying Lagrange formulation for each element allows us to recover the equations of motions of the $(6n + 18)$ degrees of freedom of the system:

$$M_P\ddot{q}_P + C_P\dot{q}_P + \left[K_{pP} + K_{eP}(t)\right]q_P = F_P(t) \tag{8}$$

where q_P, M_P, C_P, K_{pP}, K_{eP}, F_P are respectively the displacement vector, the mass, the damping, the bearing, the mesh stiffness and the force matrices for the planetary gear stage.

$$q_{PG} = \left\{ \begin{array}{l} u_c, v_c, w_c, \rho_{cx}, \rho_{cy}, \rho_{cz}, u_r, v_r, w_r, \rho_{rx}, \rho_{ry}, \rho_{rz}, u_s, v_s, w_s, \rho_{sx}, \rho_{sy}, \rho_{sz}, \\ u_1, v_1, w_1, \rho_{1x}, \rho_{1y}, \rho_{1z}, \ldots, u_n, v_n, w_n, \rho_{nx}, \rho_{ny}, \rho_{nz} \end{array} \right\}^T \tag{9}$$

2.2 Spiral Bevel Gear

For the spiral bevel gear pair, the transmission is divided into two rigid blocks. Each block has four degrees of freedom three translations x_i, y_i, z_i $(i = 1, 2)$ and one rotation θ_1 for pinion, θ_2 for wheel and θ_m for the motor. The two gear bodies are considered as rigid cone disks and the shafts with torsional stiffness. The mesh stiffness is modeled by a linear stiffness acting along the line of action (Karray et al. 2012).

The vector defining the different degrees of freedom is:

$$q_{BG} = \{x_1, y_1, z_1, \theta_m, \theta_1, x_2, y_2, z_2, \theta_2\}^T \tag{10}$$

The transmission error λ can be defined by (Gosselin et al. 1995):

$$\lambda = (-x_1 + x_2)a_1 + (y_1 - y_2)a_2 + (z_1 - z_2 + r_{m1}\theta_{1x} - r_{m2}\theta_{2y})a_3 \tag{11}$$

where

$$a_1 = \sin(\alpha)\sin(\delta_1) + \cos(\alpha)\sin(\beta)\cos(\delta_1) \tag{12}$$

$$a_2 = \sin(\alpha)\cos(\delta_1) - \cos(\alpha)\sin(\beta)\sin(\delta_1) \tag{13}$$

$$a_3 = \cos(\alpha)\cos(\beta) \tag{14}$$

β is the spiral angle, δ_1 is the pinion pitch angle, α is the pressure angle and r_{m1} and r_{m2} are respectively the means radius of the pinion and the wheel.

The equation of motion of the system is obtained by applying Lagrange formulation and is given by:

$$M_{BG}\ddot{q}_{BG} + C_{BG}\ddot{q}_{BG} + K_{BG}(t)q_{BG} = F_{BG} \tag{15}$$

where q_{BG}, M_{BG}, C_{BG}, K_{BG}, F_{BG} are respectively the displacement vector, the mass, the damping, the mesh stiffness and the force matrices for the bevel gear stage.

Table 2 Parameters of the planetary gear model

	Sun	Ring	Carrier	Planet
Teeth number	30	70	–	20
Mass (Kg)	0.46	0.588	3	0.177
J/R_{bi}^2	0.272	0.795	1.5	0.1
I/R_{bi}^2	0.136	0.389	0.75	0.05
Base radius R_{bi} (m)	0.024	0.056	0.043	0.016
Module	1.7			
Gear mesh stiffness (N/m)	$k_{sp} = 2{\cdot}10^8$, $k_{rp} = 2.3{\cdot}10^8$			
Bearing stiffness (N/m)	$k_{jx} = k_{jy} = 10^8, j = c, s,\ k_{jz} = 10^9, j = c, s$			
	$k_{rx} = k_{ry} = k_{rz} = 10^{10},\ k_{xx} = k_{yy} = 10^8,\ k_{zz} = 10^9$			
Torsional stiffness (N/m)	$k_{j\varphi} = k_{j\psi} = 10^9, j = c, s, 1\ldots n\ , k_{j\theta} = 0, j = c, s, 1\ldots n$			
	$k_{r\varphi} = k_{r\psi} = k_{r\theta} = 10^{10}$			
Pressure angle	$\alpha = 21.34°$			
Helix angle	$\beta = 20°$			

Table 3 Parameters of the bevel gear model

Parameters	Pinion	Gear
Number of teeth Z	17	27
Pressure angle α		20
Mass (kg)	0.5	1.5
Moment of inertia (kg·m^2)	0.2	0.4
Axial stiffness k_{x1}, k_{y2} (N/m)	$1{\cdot}10^9$	$2.3{\cdot}10^9$
Lateral stiffness k_{y1}, k_{z1}, k_{x2}, k_{z2} (N/m)	$8.8{\cdot}10^9$	$1.3{\cdot}10^{10}$
Torsional stiffness $k_{\theta 1}$, $k_{\theta 2}$ (Nm/rd)	$1.2{\cdot}10^4$	$7.4{\cdot}10^4$

2.3 Coupling of the Models and Eigenvalue Problem

Coupling between the planetary gear system and the bevel gear is done using an additional torsional stiffness joining rotational degree of freedom of wheel and sun gear and a linear spring joining axial degrees of freedom of wheel and sun.

The equations of motions of the (6n + 18 + 9) degrees of freedom of the system is:

$$M_G\, \ddot{q}_G + C_G\, \ddot{q}_G + \left[K_{pG} + K_{eG}(t)\right]\, q_G \;=\; F_G(t) \tag{16}$$

The undamped eigenvalue problem derived from the equation of motion by considering only the mean stiffness matrix $\overline{K}$ is:

$$\left(-\omega_i^2 \mathrm{M} + \overline{\mathrm{K}}\right) \phi_i = 0$$

where ϕ_i is the eigenvector and ω_i the corresponding eigenfrequency.

Table 4 Eigenfrequencies of the system

Mode type	Eigenfrequencies (Hz)
Coupled mode	$f_1 = 0, f_2 = 37, f_3 = 57, f_4 = 359, f_{10} = 3103, f_{12} = 3687, f_{15} = 3950, f_{22} = 8010, f_{23} = 8750,$ $f_{27} = 11808, f_{29} = 14019, f_{32} = 14817$
Bevel gear mode	$f_{18} = 7119, f_{33} = 14866, f_{40} = 21114, f_{41} = 21243$
Planetary mode (first class)	$f_7 = 2575, f_{36} = 18492, f_{37} = 20840, f_{45} = 22740, f_{49} = 24478$
Planetary mode (second class)	$f_5 = f_6 = 1176, f_8 = f_9 = 2588, f_{13} = f_{14} = 3904, f_{16} = f_{17} = 4601, f_{20} = f_{21} = 7760,$ $f_{25} = f_{26} = 10090, f_{30} = f_{31} = 14190, f_{34} = f_{35} = 15253, f_{38} = f_{39} = 21009, f_{43} = f_{44} = 22720,$ $f_{47} = f_{48} = 24300, f_{50} = f_{51} = 25700$
Planetary mode (third class)	$f_{11} = 3627, f_{19} = 7481, f_{24} = 9423, f_{28} = 12677, f_{42} = 22699, f_{46} = 24224$

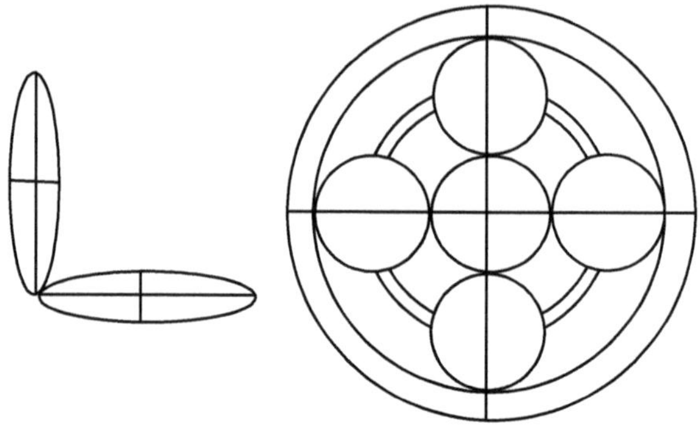

Fig. 2 Reference position

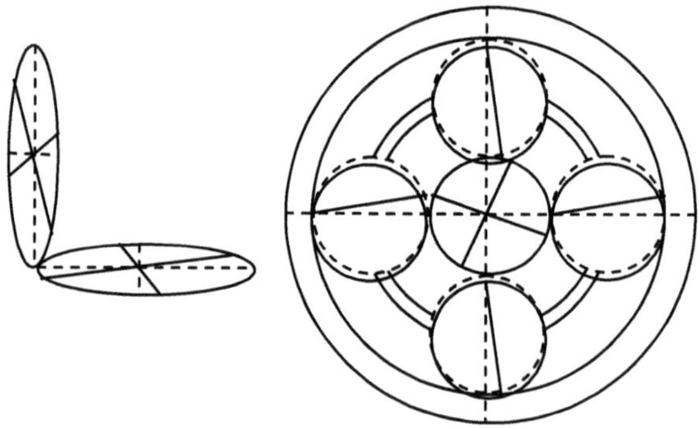

Fig. 3 Mode shapes examples for coupled mode (f = 37 Hz)

3 Modal Analysis

The studied planetary gear has a fixed ring and four planets; it is presented on Table 2 while bevel gear is presented on Table 3. The additional torsional stiffness is about $7{\cdot}10^4$ Nm/rd while the linear one is about $1{\cdot}10^9$ N/m.

The gyroscopic effect is neglected ($\mathbf{G} = \mathbf{0}$). Eigenfrequencies are presented on Table 4. The natural modes can be classified into three major classes:

Figures 3, 4 and 5 shows the modal deflections with respect to the reference position (Fig. 2).

- The coupled class which includes 12 distinct natural frequencies. In this class there is a movement of both coupled system (planetary gear and bevel gear).
- The bevel gear class which includes 4 distinct natural frequencies.
- The planetary class which includes three different classes:
  - The first class which includes 5 distinct natural frequencies. The carrier, ring and sun rotate around $\overrightarrow{z}$ and translate along the same axe. The planets move identically and in phase (Fig. 4).

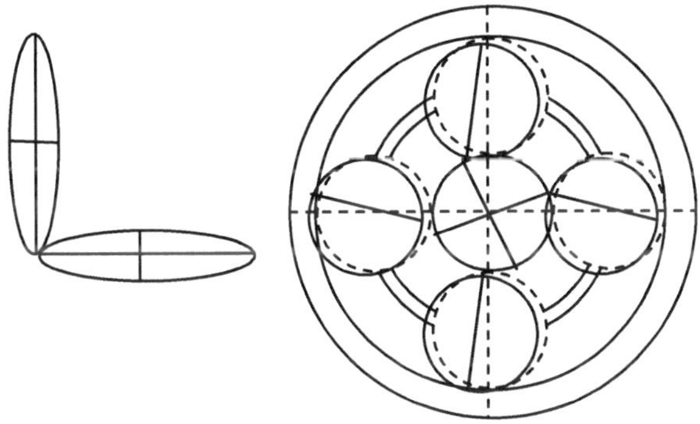

Fig. 4 Mode shapes examples for planetary mode (f = 2575 Hz)

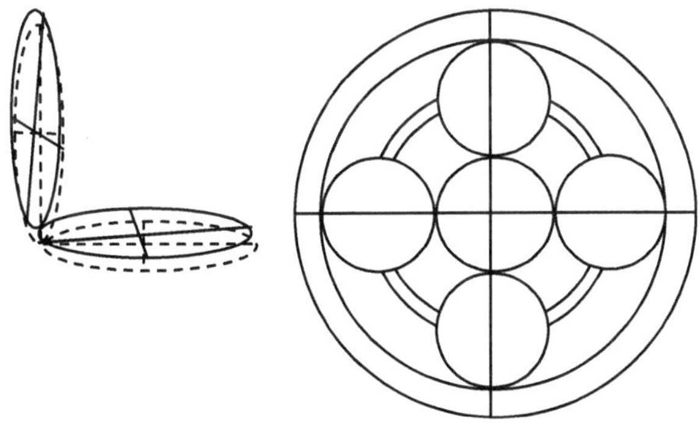

Fig. 5 Mode shapes examples for bevel gear mode (f = 14866 Hz)

- The second class includes 12 natural frequencies with multiplicity (m = 2). The carrier, ring and sun rotate and translate around $\overrightarrow{s_1}$ and $\overrightarrow{t_1}$.
- The Third class includes 6 distinct natural frequencies. Only planet motion is observed and they are counter-phased.

4 Conclusion

In this chapter a system composed by planetary gear train connected to bevel gear by an additional torsional stiffness joining rotational degree of freedom of wheel and sun gear and a linear spring joining axial degrees of freedom of wheel and sun is developed. In this model 6 degrees of freedom of planetary gear and 4 degrees of freedom of bevel are considered. The equation of motion of (6 × n + 18 + 9) has been derived in order to determine the natural frequencies of the system. It has been show that the natural mode can be classified into 3 major classes according to their displacements and rotations. It can be a bevel gear mode or a planetary mode or a coupled one.

References

Bartelmus W (2011) Gearbox damage process. In: 9th international conference on damage assessment of structures

Bartelmus W, Zimroz R (2011) Vibration spectra characteristic frequencies for condition monitoring of mining machinery compound and complex gearboxes. Scientific papers of the institute of Mining, vol 133. Wroclaw University of Technology, Poland, pp 17–34

Gosselin C et al (1995) A general formulation for the calculation of the load sharing and transmission error under load of spiral bevel and hypoid gears. Mech Mach Theory 30:433–450

Hbaieb R, Chaari F, Fakhfakh T, Haddar M (2005) Influence of eccentricity, profile error and tooth pitting on helical planetary gear vibration. J Mach Dyn Prob 29:5–32

Hidaka T, Terauchi Y, Fuji M (1980) Analysis of dynamic tooth load on planetary gear. Bull JSME 23:315–323

Kahraman A (1994a) Dynamics of a multi-mesh helical gear train. J Mech Des 116:706–712

Kahraman A (1994b) Planetary gear train dynamics. ASME J Mech Des 116:713–720

Karray M, Chaari F, Viadero F, Fdez. del Rincon A, Haddar M (2012) Dynamic response of single stage bevel gear transmission in presence of local damage. In: 4th European conference on mechanism science, EUCOMES'2012, 18–22 Sept 2012, Santander (Spain). New trends in mechanism and machine science, mechanisms and machine science, vol 7, 2013, pp 337–345. doi:10.1007/978-94-007-4902-336

Li Y et al (2010) Influence of asymmetric mesh stiffness on dynamics of spiral bevel gear transmission system. Math Prob Eng, 2010:13 (article ID 124148)

Li J et al (1998) Static analysis of bevel gears using finite element method. Commun Numer Methods Eng 14:367–380

Lin J, Parker RG (1999) Analytical characterization of the unique properties of planetary gear free vibration. J Vib Acoust 121:316–321

Maatar M, Velex P (1995) An analytical expression for the time varying contact length in perfect cylindrical gears: some possible applications in gear dynamics. J Mech Des 118:586–589

Ozguven HN, Houser DR (1988a) Dynamic analysis of high speed gears by using loaded static transmission error. J Sound Vib 125:71–83

Ozguven HN, Houser DR (1988b) Mathematical models used in gear dynamics—a review. J Sound Vib 121:383–411

Dynamic Characterization of Viscoelastic Components

Hanen Jrad, Jean Luc Dion, Franck Renaud, Imad Tawfiq and Mohamed Haddar

Abstract Characterizing frictional behavior of viscoelastic joints is investigated in the present work. A new visco-tribological model was developed by coupling the rheological Generalized Maxwell model (GMM) and Dahl friction model. Parameters of the proposed model are identified from Dynamic Mechanical Analysis (DMA) tests for different excitation frequencies. Comparison between measurements and simulations of hysteretic friction of the viscoelastic component has been carried on.

Keywords Vibration damping · Generalized Maxwell model · Dynamic friction · Dahl model

H. Jrad (✉) · J. L. Dion · F. Renaud · I. Tawfiq
Laboratoire d'Ingénierie des Systèmes Mécaniques et des Matériaux (LISMMA), Institut Supérieur de Mécanique de Paris, 3 rue Fernand Hainaut, 93407 Saint Ouen Cedex, Paris, France
e-mail: hanen.j@gmail.com

J. L. Dion
e-mail: jean-luc.dion@supmeca.fr

F. Renaud
e-mail: franck.renaud@supmeca.fr

I. Tawfiq
e-mail: imad.tawfiq@supmeca.fr

H. Jrad · M. Haddar
Laboratoire Modélisation, Mécanique et Productique (LA2MP), Ecole Nationale d'Ingénieurs de Sfax, BP N° 1173, 3038 Sfax, Tunisie
e-mail: mohamed.haddar@enis.rnu.tn

M. S. Abbes et al. (eds.), *Mechatronic Systems: Theory and Applications*,
Lecture Notes in Mechanical Engineering, DOI: 10.1007/978-3-319-07170-1_15,

1 Introduction

This chapter focuses on both internal and interfaces damping of viscoelastic joints. In most of Multi-body simulations (MBS), the dynamic behavior of viscoelastic joints is usually overlooked or, at best, greatly simplified. Hence, to improve description and prediction, global models of mechanisms must represent joints between solids, not only to model the free relative degrees of freedom to take into account the kinematics involved but also stiffness and damping in the broad meaning.

Several experimental studies have been carried out to characterize dynamic behavior of viscoelastic components depending on frequency. Oberst and Frankenfeld (1952), Barbosa and Farage (2008), Castello et al. (2008) and Chevalier (2002) use Frequency Response Function (FRF) which can only characterize the frequencies of modes and not on a wide frequency band. Chen (2000) suggested measuring directly the relaxation functions and creep to deduce the coefficients of a series of Prony. But to get high frequency values, a perfect unit step function is required to assess when exciting the material, which is technically difficult to achieve. In this work, Dynamic Mechanical Analysis (DMA) is used to determine the dynamic characteristics of the viscoelastic component depending on the frequency.

Furthermore, different models describing viscoelastic behavior have been developed. Gaul et al. (1991) presented the constant complex stiffness modulus model which is non-causal model, it is only suitable in the frequency domain, but, it is not a relevant model since its magnitude is constant. Maxwell model represented by Park (2001) as a spring and dash-pot connected in series and Kevin Voigt model which consists of a spring and dash-pot in parallel, are efficient only on a small frequency range. In fact, they are unrealistic respectively at low and high frequencies, where their magnitudes are respectively: infinitely small and high and the dynamic stiffness phase angle of the Kevin Voigt model is linearly dependent of frequency. The Zener model, see Huynh et al. (2002), underestimates the dynamic stiffness at low frequencies and overestimates it at high frequencies. Just as the Kevin Voigt model, the Zener model is unable to capture the frequency dependence of the phase angle. Koeller (1984) used Generalized Maxwell Model (GMM), which would refer to a spring in parallel with respectively Maxwell cells, to describe the frequency dependence of the dynamic stiffness of the viscoelastic components. In this study, GMM is used to characterize dynamic behavior of viscoelastic joints.

Energy losses are not only due to viscoelastic nature of joints but it is also strongly linked to the friction properties of joints. In fact, friction forces at clamping areas at joints play an important role in the energy dissipations. Numerous models of friction have been developed and presented in the literature. Tangential friction forces are often modeled by a Coulomb friction force. Coulomb model describes the steady-state behavior between velocity and friction force and does not specify the friction force for zero velocity. This discontinuity does not reflect the real friction behavior in a good way and may lead to unwanted numerical problems. Karnopp in (1985) developed a model to overcome this problem, but, his model is so strongly coupled with the rest of the system. The external force is an input to the model and this force is not always explicitly given. Moreover, experiments have shown that friction exhibits phenomena that cannot be modeled with static friction models, such as presliding displacement, and stick-slip motion. The seven parameter model proposed by Armstrong in (1994) include relevant experimentally observed friction phenomena. Although useful for analysis of stick-slip behavior, for simulation purposes the model seems to be less appropriate, see Eborn and Olsson (1995). The seven parameter model and similar static models try to capture the dynamics of friction by introducing time dependency or a time delay. To improve the description of dynamic behavior, other models have been developed with internal state variables that determine the level of friction and velocity. The evolution in time of the state variables is governed by a set of differential equations. The Dahl Model (1968) incorporates a state variable to model presliding displacement. Dahl starting point was the linear stress-strain curve and steady-state version of the Dahl model is Coulomb friction. Haessig and Friedland (1991) assumed friction between two contact surfaces to be caused by a large number of interacting bristles and introduced the bristle model which capture the randomness of friction that originate from the random distribution of asperities on a surface. The bristle model gives interpretation of friction as interacting bristles and reproduces a real-life random behavior. However, numerically it is highly ineffective and not well used in simulations. The reset integrator model introduced also by Haessig and Friedland (1991) is numerically more efficient for simulation purposes which gather the dynamic effects of friction using an integrator with a reset action to distinguish between the two cases sticking and slipping. This model is discontinuous in the state variable due to switching from sticking to sliding and numerical problems may arise for very large damping values or spring stiffness. The LuGre model suggested by Canudas de Wit et al. (1995), was designed to extend Dahl's model to include other dynamic friction effects, such as those associated with the sliding of lubricated contacts. Dahl model is used in this study to characterize friction of viscoelastic joints.

In this chapter, after having presented a description of the experimental procedure to characterize dynamics of the viscoelastic joint, the proposed visco-tribological model is depicted in Sect. 3. In Sect. 4, Identification techniques of the viscoelastic and tribological parameters of the suggested model are detailed and comparison between identified and measured values is also performed in order to investigate the validity of the proposed model.

2 Experimental Characterization

2.1 Theoretical Basis

When a material is subjected to a sinusoidal cyclic displacement of angular frequency ω:

$$x(t) = x_0 \sin(\omega t) \tag{1}$$

The force response is sinusoidal at the same frequency but with a dephasing angle φ, called loss angle:

$$F(t) = F_0 \sin(\omega t + \varphi) \tag{2}$$

Generally, this assumption, called the first harmonic, is not sufficient. Typically, the force response contains higher order harmonics, and the real response is expressed as follows, Long (2005):

$$F(t) = \sum_k F_k \sin(k\omega t + \varphi_k) \tag{3}$$

Figure 1 corresponds to linear viscoelastic behavior, which is characterized by a pure elliptical form. In the case of the assumption of the first harmonic, the energy dissipated as heat during a cycle corresponds to the area enclosed by the loading and unloading curves shown in Fig. 1 and is expressed as follows:

$$\begin{aligned} E_d &= \oint F(t)dx = F_0 x_0 \int_0^T \cos(\omega t)\sin(\omega t + \varphi)dt \\ &= \pi F_0 x_0 \sin(\varphi) \end{aligned} \tag{4}$$

The dissipated energy during a hysteresis cycle as heat reflects the damping properties of the material. The complex stiffness $K^*(\omega)$ relates the Fourier transform of the imposed displacement $\hat{x}(\omega)$ to the corresponding force $\hat{F}(\omega)$ is defined as follows:

$$\hat{F}(\omega) = K^*(\omega)\,\hat{x}(\omega) \tag{5}$$

with the Fourier transform :

$$\hat{x}(\omega) = \int_{-\infty}^{+\infty} x(t)\ \exp(-j\omega t)\,dt \tag{6}$$

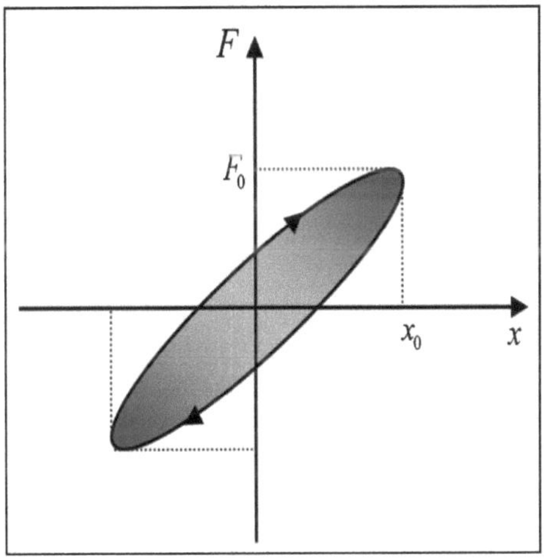

Fig. 1 Hysteresis loop for harmonic excitation

$$\hat{F}(\omega) = \int_{-\infty}^{+\infty} F(t)\ \exp(-j\omega t)\,dt \tag{7}$$

The dynamic stiffness is defined as:

$$\begin{aligned} K^*(\omega) &= \frac{\hat{F}(\omega)}{\hat{x}(\omega)} = \frac{F_0}{x_0}.\exp(j\varphi) \\ &= K'(\omega) + jK''(\omega) \\ &= K'(\omega)[1 + j\tan\varphi] \end{aligned} \tag{8}$$

where the term K' is the real part of the dynamic stiffness and and K'' its imaginary part.

2.2 Experiment

The DMA tester is a bench test performed to characterize the behavior of dynamic shear of a rubber sample as shown in Fig. 2. The cylindrical elastomeric sample is subjected to shear tests under preload. The mechanical solicitation is performed using a hydraulic cylinder with an Linear Variable Differential Transformer (LVDT) displacement sensor. The system is also equipped with a force sensor built into the base of the assembly apparatus. The force and displacement signals after analog conditioning are returned on a spectrum analyzer for digital processing. The entire system is controlled by a computer equipped with an interface General Purpose Interface Bus (GPIB) card connected to the Fast Fourier Transform (FFT)

Fig. 2 The bench test set up

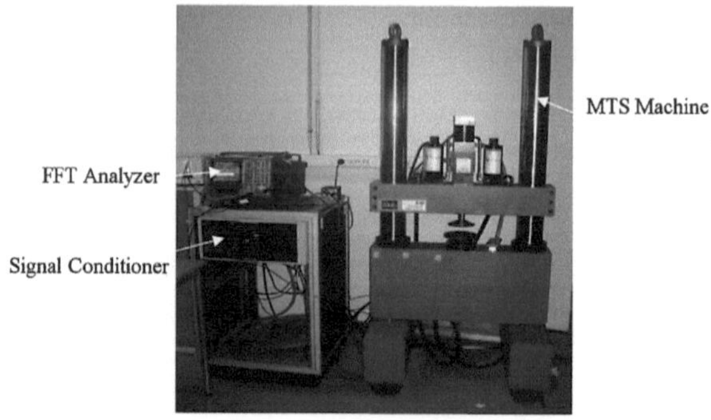

Fig. 3 Shear test

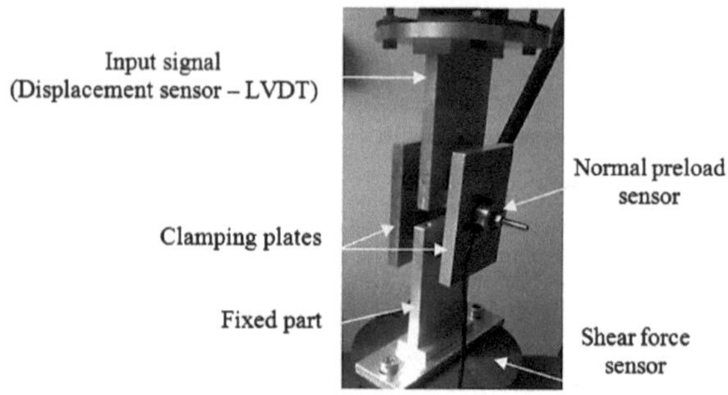

analyzer which manages and controls the sweeping frequencies of the signal by incrementing the excitation frequency for each acquisition.

To determine the dynamic stiffness K^* of the material, a double shear assembly is used as shown in Fig. 3a and b. The sample is placed between two rigid surfaces. Surfaces are flat and parallel. The two external frames are fixed when a movement is imposed on the central frame. Assuming that the deformation is homogeneous, which means there is no deformation other than the horizontal one.

During tests, devices measure the generated force F and the horizontal displacement x of the upper surface of the sample. For a cylindrical rubber sample of height $H_0 = 18$ mm, and diameter $D_0 = 26$ mm, the surface on which the force acts is $S_0 = \pi \times (D_0/2)^2$. The dynamic stiffness is given by:

$$K^* = \frac{F}{x} \tag{9}$$

These tests are carried out to evaluate the dynamic behavior of elastomeric sample and are performed by applying a mechanical sinusoidal solicitation. Elastomeric materials present a different behavior according to amplitude of sinusoidal displacement and conforming to the imposed static preload [Soula and Chevalier (1998), Huynh et al. (2002), Moreau (2007) and Saad (2003)]. Tests are carried out for different static preloads $P = [150, 500, 1000]$ N. Measuring devices

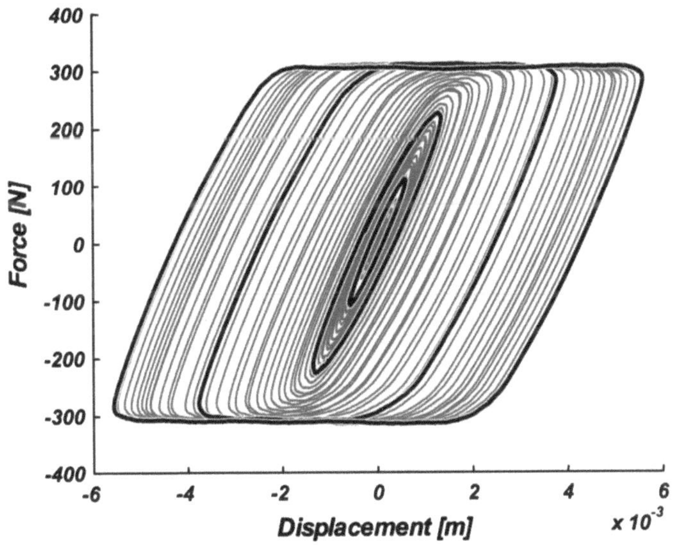

Fig. 4 Force-displacement relation for rubber component submitted to cyclic double shear loading

allow calculating the dynamic shear modulus for amplitudes ranging to 5 mm and for frequencies ranging from 2.5 to 70 Hz. Measures were taken when cycles were stable and the stability of the cycle is expected to avoid taking into account the disturbance due to transitory states. Experiments were performed at room temperature $T = 20\,^{\circ}C$.

2.3 Results

Force and displacement are used here to characterize the dynamic behavior of the viscoelastic component. The diagram in Fig. 4 shows the evolution of tangential force F versus displacement x. The first cycles of the test show ellipsoid curves which correspond to viscoelastic behavior. The non-linear behavior appears as a distortion of the pure elliptical form of the curves. These non-linearities are due to frictional damping. For small displacements, friction is driven by micro-slip and partial slip, while for high displacement; the shape of the curve clearly shows macro slipping behavior. The sticking phases are highlighted by the vertical section of the curve: macro sliding is no longer observed and the entire displacement measured is due to elasticity. The quasi vertical part of the curve represents the system stiffness. The horizontal part of the curve is the significant behavior of the slip. Most parametric models Dahl (1968), Canudas de Wit et al. (1995), Al Majid and Dufour (2004), Awrejcewicz (2003) can be identified with this kind of diagram.

3 Description of Hysteretic Friction Model

3.1 The Dahl Model

Most dynamic models of frictional behavior are built with internal state variables of kinematic type; however these variables cannot be identified as real physical displacements [Dahl (1968), Canudas de Wit et al. (1995), Chevallier (2005) and Stribeck (1902)]. The first model of this kind was proposed by Dahl (1968). Dahl proposed to model friction by the following relation:

$$\begin{cases} \dfrac{dF}{dx} = \sigma \left| 1 - \dfrac{F}{F_{\max}} sign(v) \right|^{\alpha} sign\left(1 - \dfrac{F}{F_{\max}} sign(v)\right) \\ F(t) = \sigma z(t) \end{cases} \tag{10}$$

where F is the frictional force, z is the internal displacement state variable, σ is homogeneous with stiffness, $F_{\max}$ the Coulomb force (maximum friction force reached) and $\alpha \in R_{*}^{+}$ is a constant parameter, x is the relative displacement and $\dot{x} = v$ the relative velocity between two contact surfaces. It is possible to rewrite the differential equation of the model with respect to the temporal variable:

$$\frac{dz}{dt} = \frac{dz}{dx}\frac{dx}{dt} \tag{11}$$

The system of equations becomes:

$$\begin{cases} \dfrac{dz}{dt} = v \left| 1 - \dfrac{F}{F_{\max}} sign(v) \right|^{\alpha} sign\left(1 - \dfrac{F}{F_{\max}} sign(v)\right) \\ F(t) = \sigma z(t) \end{cases} \tag{12}$$

Figure 5 illustrate hysteresis loop of Dahl model. Frictional force F increases asymptotically toward $F_{\max}$ for $\dot{x} \succ 0$ and toward $F_{\max}$ for $\dot{x} \prec 0$. The rising and downward curves are not merged, it creates the hysteresis. σ is the slope of the force-displacement curve when $F = 0$.

3.2 The Generalized Maxwell Model

GMM allows an accurate description of the dynamic behavior of a viscoelastic joint. GMM is classically composed of Maxwell cells in parallel. A Maxwell cell is represented by a spring and dash-pot connected in series. With such definition this model is not able to display reversible creep, see Caputo and Mainardi (1971). As this chapter deals only with viscoelastic solids, GMM would refer to a spring in

Fig. 5 Hysteresis loop of Dahl model

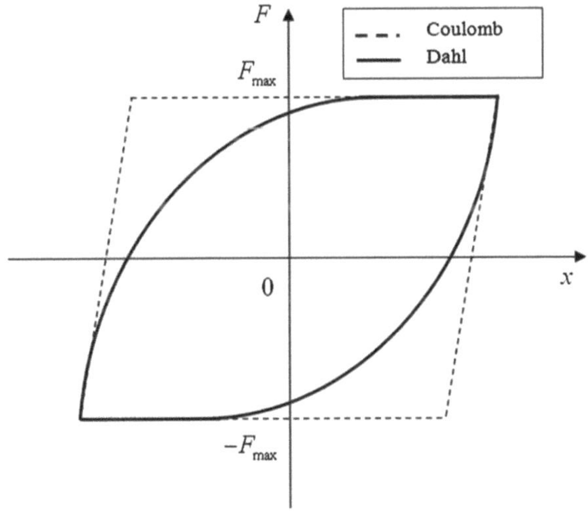

parallel with respectively Maxwell cells, see Koeller (1984). Thus, the GMM defined here is the same as that used by Chevalier and Vinh (2010) and the same as the Maxwell representation given by Caputo and Mainardi (1971), without the first dashpot.

To model the non-linear dynamic behavior of the viscoelastic joint, the chosen linear GMM is composed of a linear spring and N linear Maxwell cells as represented in Fig. 6.

To deform this rheological model, it is necessary to impose a displacement $x(t)$, the response is the sum of the spring force added to each cell reaction, noted $F(t)$:

$$F(t) = F_0(t) + \sum_{i=1}^{N} F_i(t) \tag{13}$$

The rheological formulation of the dynamic stiffness of GMM is

$$Z(\omega) = K_0 + \sum_{i=1}^{N} \frac{j\omega K_i C_i}{K_i + j\omega C_i} \tag{14}$$

K_0 is the stiffness taken at $\omega = 0$, i.e., $t = +\infty$, K_i is the stiffness of the ith spring and C_i is the damping of the ith dashpot. Reducing Eq. (14) to the same denominator and grouping monomials gives the dynamic stiffness of linear GMM expressed as the ratio of two polynomials of the same degree N (number of Maxwell cells). This formulation of transfer function is also used in automation, namely, Oustaloup (1991) provided a model using poles and zeros formulation (PZF).

Fig. 6 The proposed linear GMM

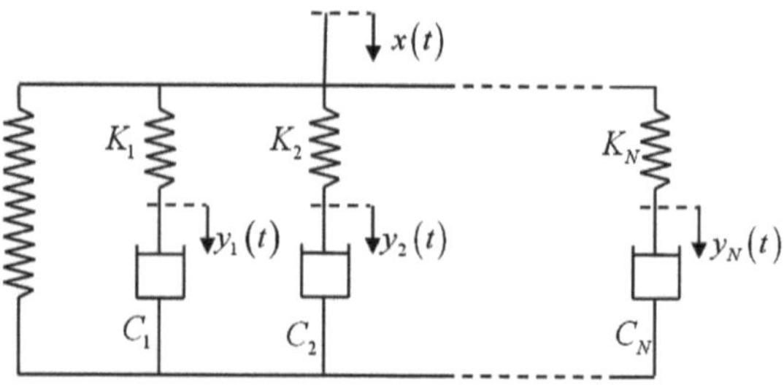

$$Z(\omega) = K_0 \prod_{i=1}^{N} \left(\frac{1 + (j\omega/\omega_{z,i})}{1 + (j\omega/\omega_{p,i})} \right) \tag{15}$$

$\omega_{z,i}$ and $\omega_{p,i}$ are respectively the zero and the pole of the ith Pole–Zero couple, $i \in [1..N]$. This operator called by Oustaloup (1991) "CRONE regulator" facilitates considerably the treatment and the parametric identification of the polynomial ratio by expressing it in the form of products.

3.3 The Visco-tribological Model

The visco-tribological model chosen is based on both of Dahl model and linear GMM model. These two models are combined by supposing that the force generated by the frictional spring of Dahl, $F(t) = \sigma z(t)$ is viscoelastic, that is to say that σ will be modeled by generalized Maxwell. During the imposed displacement $x(t)$, the frictional spring will stretch of $z(t)$. It is considered that $z(t) = x(t)$.

$$\begin{cases} \dot{z}(t) = \dot{x}(t)\left|1 - \dfrac{F(t)}{F_{\max}} sign(\dot{x}(t))\right|^{\alpha} sign\left(1 - \dfrac{F(t)}{F_{\max}} sign(\dot{x}(t))\right) \\ F(t) = K_0 z(t) + \displaystyle\sum_{i=1}^{N} K_i(z(t) - y_i(t)) \\ K_i(z(t) - y_i(t)) = C_i \dot{y}_i(t) \\ F_{\max} = \mu F_N \end{cases} \tag{16}$$

where μ is the friction coefficient and F_N is the normal load.

4 Parametric Identification Techniques

The hysteresis loops of force-displacement contain informations which enable identification of parameters of the proposed model.

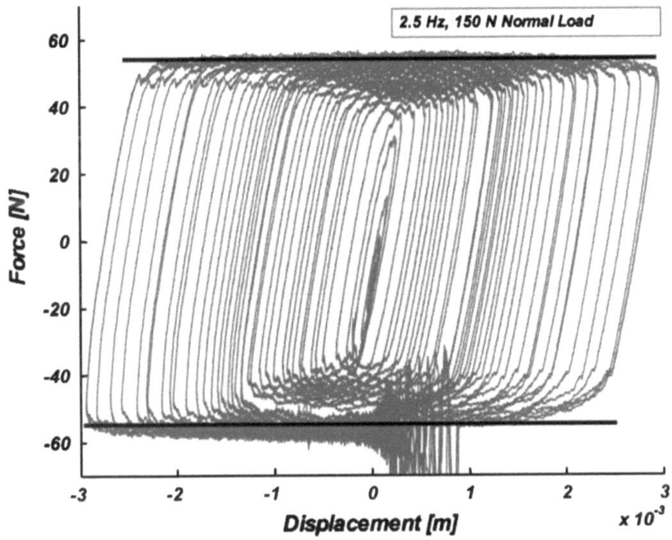

Fig. 7 Friction coefficient identification

4.1 Friction Coefficient Identification

The friction coefficient μ is defined by the horizontal tangent line when F reaches its maximum as shown in Fig. 7.

$$F_{\max} = \mu P \tag{17}$$

The friction coefficient identified value for all tests is $\mu = 0.3$.

4.2 Viscoelastic Parameters Identification

4.2.1 K_0 Identification

Force-displacement relationships present Hysteresis loops. The dynamic stiffness is given by the slope of the hysteresis loop. The first cycles of the diagram show pure ellipsoid curves which correspond to viscoelastic behavior. The quasi vertical part of the curve represents the system stiffness. The dynamic stiffness value is estimated to be the slope of the major axis of an ellipse which coincides with one selected cycle of the first cycles as illustrated in Fig. 8.

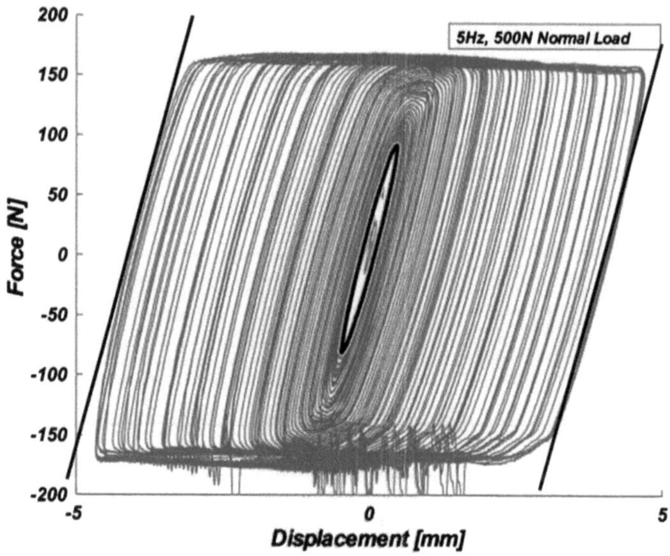

Fig. 8 K_0 identification

4.2.2 K_i and C_i Identification

Renaud (2011) demonstrated relations (18) which allow computing NLGMM parameters, given by Eq. (14), from the parameters of PZF (Poles-Zeros Formulation).

$$\begin{cases} K_i = K_0 \prod_{h=1}^{N} \left(\frac{\omega_{p,h}}{\omega_{z,h}}\right) \left(\frac{\omega_{p,i} - \omega_{z,h}}{\omega_{p,i} + \omega_{p,h}(\delta_{ih} - 1)}\right) & (a) \\ C_i = \frac{K_i}{\omega_{p,i}} & (b) \end{cases} \tag{18}$$

Considering $N = 4$ Maxwell cells, K_i and C_i coefficients are identified through Eq. (18a) and (18b), after computing the poles and zeros.

The method used to determine the poles and zeros is analogous to the one proposed by Oustaloup (1991). The main idea of this approach leads to consider the angle phase equal to $\pi/2$ between zero and pole of the same order and null elsewhere. The resulting phase angle in the studied frequency range is then estimated as the average of phase angle calculated between the first zero and the last pole, see Dion (1995).

To obtain a constant phase between two consecutive zeros, the ratio between two consecutive zeros is constant and equal to the ratio between two consecutive poles. Two constants λ and θ are then defined:

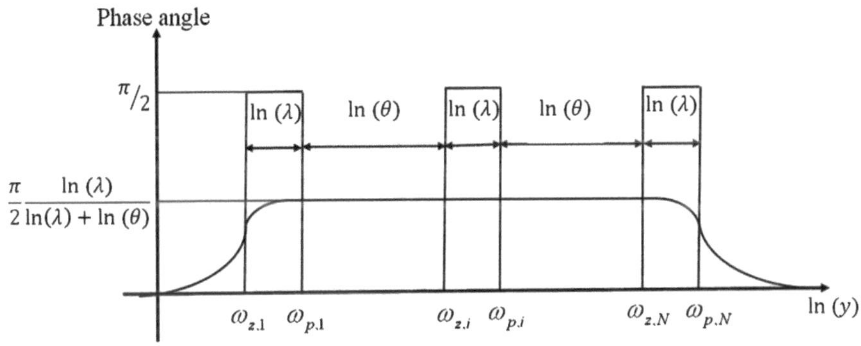

Fig. 9 Approach of constant phase angle

$$\ln(\lambda) = \phi \frac{\ln(f_2) - \ln(f_1)}{\phi + \frac{\pi}{2}(N-1)} \tag{19}$$

$$\ln(\theta) = \left(\frac{\pi}{2} - \phi\right) \frac{\ln(f_2) - \ln(f_1)}{\phi + \frac{\pi}{2}(N-1)} \tag{20}$$

f_1 and f_2 are respectively the upper and lower bounds of the frequency domain on which ϕ the mean phase angle for all tests is identified and $\ln(y)$ is the natural logarithm of y. The first zero coincides with $f_1$and the last pole with f_2. This approach is illustrated in Fig. 9.

The identification of a viscoelastic behavior can be performed using a ratio of two polynomial functions defined by zeros and poles.

Zeros are defined from the first zero, so that

$$\omega_{z,i+1} = \omega_{z,i}\lambda\theta \tag{21}$$

Poles are calculated from the last pole, so that

$$\omega_{p,i} = \frac{\omega_{p,i+1}}{\lambda\theta} \tag{22}$$

The relation between zeros and poles is given by

$$\omega_{p,i} = \lambda\omega_{z,i} \tag{23}$$

5 Validation Results

The dynamic stiffness of viscoelastic components is well described by its magnitude and phase $Z(\omega) = |Z(\omega)|\exp(j\varphi(\omega))$. Renaud (2011) presented a method based on characteristics of the asymptotes of Pole–Zero formulations which allows

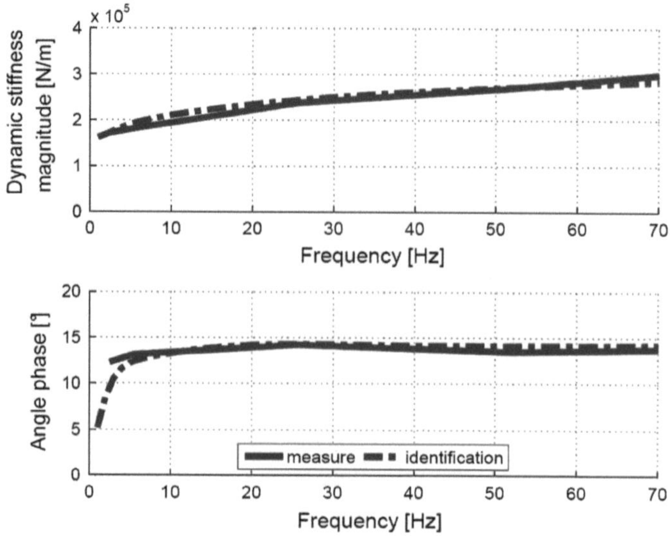

Fig. 10 Measured and identified dynamic stiffness magnitude and phase angle as function of frequency for P = 500 N

identifying GMM parameters from both the magnitude and the phase curves with more efficiency than the classical graphical methods thanks to optimization algorithm based on asymptotes. Magnitude and phase of the associated PZF are defined in Eq. (24)

$$\begin{cases} |Z(\omega)| = K_0 \prod_{i=1}^{N} |Z(\omega)|_i = K_0 \prod_{i=1}^{N} \dfrac{\sqrt{1 + (\omega/\omega_{z,i})^2}}{\sqrt{1 + (\omega/\omega_{p,i})^2}} \\ \varphi(\omega) = \sum_{i=1}^{N} \varphi_i(\omega) = \sum_{i=1}^{N} \left(\tan^{-1}\left(\dfrac{\omega}{\omega_{z,i}}\right) - \tan^{-1}\left(\dfrac{\omega}{\omega_{p,i}}\right) \right) \end{cases} \tag{24}$$

Figure 10 show good agreements between measured and identified values of magnitude and phase of dynamic stiffness of the viscoelastic component. The hardening behavior of the viscoelastic joint for increasing preloads is well identified.

The proposed visco-tribological model is validated in Fig. 11.

The identification method proposed is robust and has been applied for a very large number of tests with several frequencies and normal loads. The quality of fitting between simulations and measurements in Fig. 11 is similar for most of the tests.

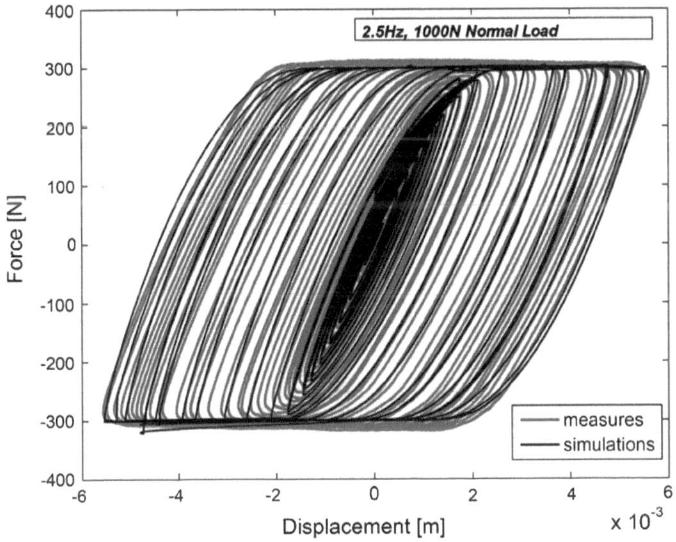

Fig. 11 Comparison between measurement and simulation obtained with the proposed identification method

6 Conclusion

The work presented here was performed in the field of structural vibrations. In a complex mechanism, the level of vibration strongly depends on its dissipation in the connected parts.

The visco-tribological model proposed in this study is an internal state variable model similar to the Dahl. This model of friction takes into account the viscoelastic dynamic behavior of the joint. An accurate method for parametric identification is performed. The model chosen is composed of one internal state variable and ten parameters (μ, K_0 and $K_i, C_i, i = [1..4]$) that can be identified with only one test. This model can be employed on several scales and implemented in structural computations.

References

Al Majid A, Dufour R (2004) Harmonic response of a structure mounted on an isolator modelled with a hysteretic operator: experiments and prediction. J Sound Vib 277(1–2):391–403

Armstrong-Hétlouvry B, Dupont P, Canudas de Wit C (1994) A survey of models, analysis tools and compensation methods for the control of machines with friction. Automatica 30(7):1083–1138

Awrejcewicz J, Lamarque CH (2003) Bifurcation and chaos in nonsmooth mechanical systems. World Sci Publ, Singapore

Barbosa F, Farage M (2008) A finite element model for sandwich viscoelastic beams: experimental and numerical assessment. J Sound Vib 317(1–2):91–111
Canudas de Wit C, Olsson H, Åström KJ, Lischinsky P (1995) A new model for control systems with friction. IEEE Trans Autom Control AC 40:419–425
Caputo M, Mainardi F (1971) Linear models of dissipation in anelastic solids. In: Society IP (ed) La Rivista del Nuovo Cimento (1971–1977), vol 1. Italian Physical Society, pp 161–198
Castello D, Rochinha F, Roitman N, Magluta C (2008) Constitutive parameter estimation of a viscoelastic model with internal variables. Mech Syst Sig Process 22(8):1840–1857
Chen T (2000) Determining a prony series for a viscoelastic material from time varying strain data, Technical report, NASA
Chevalier Y (2002) Essais dynamiques sur composites. caractérisation aux basses fréquences, Technical report, Techniques de l'ingénieur
Chevalier Y, Vinh JT (2010) Mechanics of viscoelastic materials and wave dispersion, vol 1. ISTE and John Wiley & Sons, Hoboken
Chevallier G (2005) Etude des vibrations de broutement provoquées par le frottement sec—application aux systèmes d'embrayage, Ph.D. thesis P6—LISMMA SUPMECA
Dahl PR (1968) A solid friction model, The Aerospace Corporation, El-Segundo, TOR-158(3107-18), California
Dion JL (1995) Modélisation et identification du comportement dynamique de liaisons hydro-élastiques, Ph.D. thesis, ISMCM, Saint-Ouen
Eborn J, Olsson M (1995) Modelling and simulation of an industrial control loop with friction. In: Proceedings of the 4th IEEE conference on control applications, Albany, New York, pp 316–322
Gaul L, Klein P, Kemple S (1991) Damping description involving fractional operators. Mech Syst Sig Process 5(2):81–88
Haessig DA Jr, Friedland B (1991) On the modeling and simulation of friction. J Dyn Syst Meas Control 113:354–362
Huynh A, Argoul P, Point N, Dion JL, (2002) Rheological models using fractional derivatives for linear viscoelastic materials: application to identification of the behavior of elastomers. In: 13th symposium noise, shock and vibration, Lyon
Karnopp D (1985) Computer simulation of slip-stick friction in mechanical dynamic systems. J Dyn Sys Meas Contr 107(1):100–103
Koeller RC (1984) Applications of fractional calculus to the theory of viscoelasticity. J Appl Mech 51(2):299–307
Long AHK (2005) Analysis of the dynamic behavior of an elastomer: modeling and identification, Ph.D. thesis, LISMMA, SUPMECA, Paris
Moreau A (2007) Identification of viscoelastic properties of polymeric materials by field measurements of frequency response of structures, Ph.D. thesis, LMR, INSA, Rouen
Oberst H, Frankenfeld K (1952) Damping of the bending vibrations of thin laminated metal beams connected through adherent layer. Acustica 2:181–194
Oustaloup A (1991) La commande CRONE: commande robuste d'ordre non entier, Hermés
Park SW (2001) Analytical modeling of viscoelastic dampers for structural and vibration control. Int J Solids Struct 38:8065–8092
Renaud F, Dion JL, Chevallier G, Tawfiq I, Lemaire R (2011) A new identification method of viscoelastic behavior: application to the generalized Maxwell model. Mech Syst Sig Process 25(3):991–1010
Saad P (2003) Nonlinear behavior of rubber bush, modeling and identification, Ph.D. thesis, LTDS, Ecole Centrale De Lyon, Lyon
Soula M, Chevalier Y (1998) The fractional derivative rheological polymers—application to elastic and viscoelastic behaviors linear and nonlinear elastomers. ESAIM Proc Fract Differ Syst 5:193–204
Stribeck R (1902) Die Wesentlichen Eigenschaften der Gleit—und Rollenlager—the key qualities of sliding and roller bearings, Zeitschrift des Vereines Seutscher Ingenieure, 46, 38, pp 1342–48, 46, 39, 1432–1437

MIX
Papier aus verantwortungsvollen Quellen
Paper from responsible sources
FSC® C105338

If you have any concerns about our products,
you can contact us on
ProductSafety@springernature.com

In case Publisher is established outside the EU,
the EU authorized representative is:
Springer Nature Customer Service Center GmbH
Europaplatz 3, 69115 Heidelberg, Germany

Printed by Libri Plureos GmbH
in Hamburg, Germany